DATE DUE

MAY 19 1982		
MAY 23 1984		
NOV 07 198_		
JAN 23 1985		
OCT 15 1986		
WITHDRAWN		

DEMCO 38-297

D1466553

COMPUTABILITY THEORY
AN INTRODUCTION

ACM MONOGRAPH SERIES

*Published under the auspices of the Association for
Computing Machinery Inc.*

Editor   ROBERT L. ASHENHURST   *The University of Chicago*

A. FINERMAN (Ed.)   University Education in Computing Science, 1968
A. GINZBURG   Algebraic Theory of Automata, 1968
E. F. CODD   Cellular Automata, 1968
G. ERNST AND A. NEWELL   GPS: A Case Study in Generality and
   Problem Solving, 1969
M. A. GAVRILOV AND A. D. ZAKREVSKII (Eds.)   LYaPAS: A Programming
   Language for Logic and Coding Algorithms, 1969
THEODOR D. STERLING, EDGAR A. BERING, JR., SEYMOUR V. POLLACK,
   AND HERBERT VAUGHAN, JR. (Eds.)   Visual Prosthesis:
   The Interdisciplinary Dialogue, 1971
JOHN R. RICE (Ed.)   Mathematical Software, 1971
ELLIOTT I. ORGANICK   Computer System Organization: The B5700/B6700
   Series, 1973
NEIL D. JONES   Computability Theory: An Introduction, 1973

In preparation
ARTO SALOMAA   Formal Languages
HARVEY ABRAMSON   Theory and Application of a Bottom-Up Syntax-
   Directed Translator

*Previously published and available from The Macmillan Company,
New York City*
V. KRYLOV   Approximate Calculation of Integrals (Translated by A. H.
   Stroud), 1962

COMPUTABILITY THEORY

AN INTRODUCTION

NEIL D. JONES

Computer Science Department
The Pennsylvania State University
University Park, Pennsylvania

ACADEMIC PRESS   New York and London   1973

WITHDRAWN
ITHACA COLLEGE LIBRARY

COPYRIGHT © 1973, BY ACADEMIC PRESS, INC.
ALL RIGHTS RESERVED.
NO PART OF THIS PUBLICATION MAY BE REPRODUCED OR
TRANSMITTED IN ANY FORM OR BY ANY MEANS, ELECTRONIC
OR MECHANICAL, INCLUDING PHOTOCOPY, RECORDING, OR ANY
INFORMATION STORAGE AND RETRIEVAL SYSTEM, WITHOUT
PERMISSION IN WRITING FROM THE PUBLISHER.

ACADEMIC PRESS, INC.
111 Fifth Avenue, New York, New York 10003

United Kingdom Edition published by
ACADEMIC PRESS, INC. (LONDON) LTD.
24/28 Oval Road, London NW1

LIBRARY OF CONGRESS CATALOG CARD NUMBER: 72-9331

AMS (MOS) 1970 Subject Classifications: 02F05, 02F15, 02F25, 68A10, 68A25

PRINTED IN THE UNITED STATES OF AMERICA

CONTENTS

v

IV. Decision of Predicates by Turing Machines

V. The Normal Form Theorems and Consequences

VI. Other Formulations of Computability

References

PREFACE

The purpose of this text is to introduce the major concepts, constructions, and theorems of the elementary theory of computability of recursive functions. The concept of "effective process" is introduced and emphasized early, so that a clear intuitive understanding of the effective computability of partial and total functions and the effective enumerability and decidability of sets is obtained before proceeding to the more rigorous aspects of the theory. This is followed by a formal development of the equivalence of Turing machine computability, enumerability, and decidability with other formulations of these concepts as defined by formulas of the predicate calculus, systems of recursion equations, and Post's production systems.

In a certain sense this text has just two purposes: to establish the fundamental properties of the partial recursive functions and the recursive and recursively enumerable sets; and to give mathematical evidence for the validity of the Church–Turing thesis.

This text is suitable for a one-semester advanced undergraduate or beginning graduate course. Alternately, Chapters I and II can be used on a stand-alone basis as an informal introduction to the various concepts related to effectiveness, as a part (say 20 to 30 percent) of an introductory course in mathematical machine theory. Such a course might also contain material on finite automata, push-down automata, and context-free languages, with a primary emphasis on the former.

The student is expected to have some informal experience with algorithms (such as that obtained in an introductory programming course), and to be generally comfortable with elementary discrete mathematics (basic logic, sets, relations, functions, etc.); it is *not* required that he be knowledgeable in higher-level mathematical subjects such as mathematical logic, abstract algebra, or number theory.

The basic approach is to prove the equivalence of effectiveness as defined by Turing machines with effectiveness as defined by S-rudimentary and other predicate calculus formulas. On the one hand, it is shown that the S-rudimentary formulas (e.g. Smullyan [S1]) can accurately describe the structure and computations of Turing machines; on the other hand, we show that Turing machines may be used to evaluate S-rudimentary and other formulas. As a consequence the equivalence of two very different formulations of computability is established, as well as normal form theorems (e.g., as in Kleene [K1]). The fundamental theorems concerning computability, enumerability, etc. are then derived from these results.

The Turing machine model used is similar to the Wang variant [W1], in that it has a *program* as a control unit.

Several factors were involved in the choice of the S-rudimentary predicates for Turing machine description, as opposed to the more customary choice of the primitive recursive functions. The author has observed that many of the difficulties encountered in teaching a first course in computability (e.g. from Davis [D1] and Hermes [H1]) seem to derive from arithmetization techniques which use various devices from number theory, such as properties of prime numbers and the Chinese remainder theorem. In addition the representation of an essentially nonnumeric object such as a Turing machine by means of a number often causes conceptual problems.

In this text the entire approach is *syntactic* rather than numeric; the basic objects are words rather than numbers (although of course numbers can be represented by words in a natural way). Thus the arithmetization of Turing machines, computations, etc. is replaced by simple *encoding* techniques, most of which merely involve replacement of symbols by strings of symbols, or the use of auxiliary symbols used as special markers.

The elegant methods of Smullyan (S1) are very well adapted to this approach; in fact we use many of his basic techniques, including the S-rudimentary predicates. As a side effect we get normal form theorems which are stronger than usual, because the function U and predicate

T_n are simpler than primitive recursive (in fact S-rudimentary). Another side effect which is of interest because of recent work in computational complexity is that U and T_n are computable and decidable (respectively) in time, which is a polynomial function of the length of their input words.

The true power of the syntactic approach is evidenced in Chapter VI in which the same tools are used to give straightforward proofs that systems of recursion equations on one hand, and Post canonical systems on the other, are both exactly equivalent to Turing machines.

Teaching experience with this approach to computability to date has indicated that it does in fact make computability theory more accessible to many students, particularly those in computer science.

LIST OF SPECIAL SYMBOLS

	Symbol	Meaning	First Reference
1.	iff	if and only if	15
2.	∎	marks the end of the definition or proof	22
3.	$x \in S$	x is an element of S	6
4.	$x \notin S$	x is not an element of S	6
5.	$S \subseteq T$	S is a subset of T	6
6.	$S \subsetneqq T$	S is a proper subset of T	6
7.	$\varnothing$	the empty set	7
8.	$\sharp S$	the number of elements in S	7
9.	$\{x \mid P(x)\}$	the set of x such that $P(x)$ is true	7
10.	$S \cup T$	the union of S and T	7
11.	$S \cap T$	the intersection of S and T	7
12.	$S \backslash T$	the difference of S and T	7
13.	(a, b)	an ordered pair	8

	Symbol	Meaning	First Reference
14.	$(x_1, \ldots, x_n)$	an ordered $n$tuple	9
15.	$\vec{x}_n$	abbreviation for the sequence $x_1, x_2, \ldots, x_n$	9
16.	$S \times T$	the cartesian product of S and T	8
17.	S^n	the $n$fold cartesian product of S with itself	9
18.	$S_1 \times S_2 \times \cdots \times S_n$	the cartesian product of $S_1, S_2, \ldots, S_n$	9
19.	λ	the empty word	12
20.	$|x|$	the number of symbols in a word x	12
21.	A, B	alphabets, that is, finite nonempty sets	11
22.	xy	the catenation of word x with word y	12
23.	x^n	the catenation of n copies of word x	12
24.	A^*	the set of all words over alphabet A	12
25.	$P(x_1, \ldots, x_n)$	an $n$ary predicate	13
26.	$\neg P(\vec{x}_n)$	"not P," the negation of $P(\vec{x}_n)$	14
27.	$P(\vec{x}_n) \wedge Q(\vec{x}_n)$	"P and Q," the conjunction of $P(\vec{x}_n)$ and $Q(\vec{x}_n)$	14
28.	$P(\vec{x}_n) \vee Q(\vec{x}_n)$	"P or Q," the disjunction of $P(\vec{x}_n)$ and $Q(\vec{x}_n)$	14
29.	$P(\vec{x}_n) \Rightarrow Q(\vec{x}_n)$	"P implies Q"	14
30.	$P(\vec{x}_n) \Leftrightarrow Q(\vec{x}_n)$	"P equivalent to Q"	14
31.	$P(\vec{\xi}_n)$	an explicit transform of $P(\vec{x}_n)$	15
32.	$\exists y\, P(\vec{x}_n, y)$	"for some y, $P(\vec{x}_n, y)$ is true," the existential quantification of $P(\vec{x}_n, y)$	15
33.	$\forall y\, P(\vec{x}_n, y)$	"for all y, $P(\vec{x}_n, y)$ is true," the universal quantification of $P(\vec{x}_n, y)$	16
34.	$\mu y\, Q(\vec{x}_n, y)$	the least y such that $Q(\vec{x}_n, y)$ is true	19

INTRODUCTION

Modern digital computers are capable of solving a great many complex problems, often in a surprisingly short time. A natural question is whether there are any problems which are beyond the capabilities of mechanical problem-solving methods, even if given arbitrarily large amounts of computing time. If any such intrinsically "unsolvable" problems exist, it is natural to wonder what their characteristics are, so we can avoid attempts to solve them.

In order to discuss these questions in anything but a philosophical way, it is necessary to define our terms. For example, a "problem" may be identified with a function $f(x, y)$; the problem of computing f in general is to find a mechanical procedure which, if given any two argument values (for instance, $x = 5$ and $y = 13$), will eventually produce the value of $f(x, y)$. For example if for all x, y $f(x, y)$ is the smallest prime number which exceeds $xy + y$, then f is clearly computable, at least in an intuitive sense.

Another way to state a problem is to define a set S of integers; the problem is to determine of an arbitrary integer x, whether or not x is a member of S. For example, S might be the set of all even numbers which can be expressed as the sum of two primes. Again it is intuitively

1

clear that there is a computation method which will decide whether a given x is in S.

It is, however, more difficult to define just what is meant by a "mechanical problem-solving method," because an embarrassingly rich array of problem-solving methods already exists. For example we have various disciplines of hand calculation (arithmetic in Roman or in Arabic numerals, algebraic manipulations, and so forth), and a great many mechanical calculation techniques, ranging from piles of stones, the Abacus, and desk calculators to a great diversity of digital computers.

It is the contention of the theory of computability that in fact it is possible to give a precise definition of the term "computable," in a way which is as powerful as any of the calculation methods above, and yet which does not go beyond our intuitive conceptions of what is effectively calculable. This definition is in terms of the "Turing machine," a mathematical model of a computing device which was introduced by the British logician, Alan M. Turing [T1]. The claim that the Turing machine is an accurate model of the intuitive concept of "computable" is known as the Church–Turing thesis; evidence is presented in Section II.4, and in Chapters V and VI.

Once precise definitions have been formulated for computability and the statements of problems, it is possible to formulate the following questions in a precise way, and to attempt to provide mathematical answers:

1. Are there unsolvable (or uncomputable) problems?
2. Are there problems which can be completely and accurately described in a finite way, which are unsolvable?
3. What characteristics of a problem can cause unsolvability?

In fact we will answer questions 1 and 2 affirmatively, and provide a number of answers to question 3.

Outline of the Book

Chapter I contains a review of basic terminology involving sets and functions and presents several mathematical topics which are particularly relevant to computability theory. These include alphabets, words and number notation in Section I.2, predicates in Section I.3, induction and inductive definitions in Section I.4, and countability and enumeration functions in Section I.5. The student should read through this chapter even if he is already familiar with some of the material, for such concepts as partial functions and enumeration functions are often not seen outside this field.

Chapter II contains an extensive informal introduction to computability theory, including informal versions of most of the main arguments of Chapter V. For this reason, the student who plans to cover Chapters III through V may not wish to spend much time on this chapter. Section II.1, "Implications of the Concept of Effectiveness," first discusses the concept of an effective (that is, mechanically performable) procedure; Subsections II.1.1 and II.1.2 establish (informally) the properties of effectively computable functions, and effectively decidable and enumerable sets and predicates. Subsection II.1.3 contains informal versions of the arguments which imply the existence of uncomputable functions and undecidable sets, involving the halting problem.

Section II.2 makes the concepts of computability, enumerability and decidability precise by means of the Turing machine, and contains a brief discussion of alternate formulations of the Turing machine.

Section II.3 contains a direct proof of the fact that a Turing machine cannot solve the halting problem for all Turing machines; that is, there is no Turing machine which, if given the program of another Turing machine Z and an input x to it, will decide whether Z will eventually halt when started on input x. In addition the construction of a universal Turing machine is outlined, which can simulate any other Turing machine.

Section II.4 contains an outline of the complex constructions of Chapters III and IV, and discussions of the Church–Turing thesis and the significance of the existence of uncomputable functions.

Chapter III begins with definition of the S-rudimentary predicates, and then gives a more formal description of Turing machines and their behavior. Following this, a number of predicates (the "basic simulation predicates") are defined, and the remainder of the chapter is devoted to showing that they are all S-rudimentary; consequently Turing computability can be described by this class of predicates.

In Chapter IV the converse of this is shown: that Turing machines can evaluate any S-rudimentary predicate. The Turing machine constructions are expressed in terms of flow charts, similar to those of Hermes [H1]. A final section, necessary only for Chapter VI, shows that auxiliary symbols do not increase computational power.

The main results of Chapters III and IV are expressed by the normal form theorems of Chapter V. This chapter also contains formal proofs of the fundamental theorems of elementary recursive function theory, mostly as direct consequences of the constructions of Chapters III and IV.

Chapter VI uses these same basic techniques to prove that two more, very different, formulations of computability are in fact equivalent to Turing computability. These are the general recursive functions of Kleene [K1] and the canonical systems of Post [P1].

∎ MATHEMATICAL BASIS

This chapter contains the basic mathematical concepts and terminology used in the theory of computability. The presentation is brief but self-contained. Some of the material may be a review for readers who are knowledgeable in set theory; however enough terminology is peculiar to computability theory so that every reader should at least skim through this chapter. Examples of new material may include partial functions, *m*adic number notation, predicates, and induction applied in nonnumeric contexts.

Readers who plan to cover only the informal aspects of computability theory (Chapters I and II) may wish to skip the material which relates to predicates and logical formulas. In general, the reader is encouraged to try all the exercises.

1. SETS AND FUNCTIONS

1.1 Sets

A set is informally defined to be any collection of objects. The only restriction that a collection must satisfy to be called a set is that for any object x, either x is definitely *in* the collection, or x is definitely *not* in it. If an object x is in a set S, we say that x is an **element** of S. Examples of sets:

1. The collection of all even integers;
2. The collection of all words in Webster's 7th New Collegiate Dictionary, 1965 Edition;
3. The three-element collection containing the integer 5, an ice cream cone, and the British Navy at 11:00 p.m., November 11, 1940;
4. The collection of all proper, nonnegative integer divisors of 28.

If a set contains a finite number of elements $a_1, a_2, \ldots, a_n$, it can be completely described by listing its elements, separating them by commas and enclosing the list in braces as follows: $\{a_1, a_2, \ldots, a_n\}$. For example, the fourth set above is $\{2, 4, 7, 14\}$.

The set described by $\{a_1, a_2, \ldots, a_n\}$ is completely given by the elements in the list; their order is irrelevant. If any element occurs more than once, all the occurrences but one could be removed from the list without changing the set. For example, the following all describe the same set:

$\{2, 4, 7, 14\}, \{4, 2, 14, 7\}, \{2, 4, 2, 7, 2, 14\}, \{2, 7, 7, 4, 4, 14\}.$

An infinite set may be described similarly, if there is an obvious rule for listing its elements as an infinite sequence. For example, the next three sets are well defined:

$$S = \{2, 4, 6, 8, \ldots\}$$
$$T = \{2, 4, 8, 16, \ldots\}$$
$$U = \{2, 3, 5, 7, 11, 13, 17, 19, \ldots\}.$$

Throughout these notes we will use the symbol N to denote the set of nonnegative integers $\{0, 1, 2, 3, \ldots\}$.

If x is an element of S, we write: $x \in S$; if x is not in S, we write $x \notin S$.

Let S and T be sets. We say that S is **contained** in T, or S is a **subset** of T, if every element of S is also an element of T. This is written: $S \subseteq T$. It should be clear that S equals T just in case $S \subseteq T$ and $T \subseteq S$ are both true. If $S \subseteq T$ and $S \neq T$, we say that S is a **proper subset** of T, written $S \subsetneqq T$. Alternate notations of $S \subseteq T$ and $S \subsetneqq T$ are $T \supseteq S$ and $T \supsetneqq S$, respectively.

Note: Some texts use $\subseteq$ to indicate containment and $\subset$ to indicate proper containment; others use $\subset$ to indicate either type of containment. To avoid ambiguity, we use the hybrid notation just given.

The set which contains no element is denoted by $\varnothing$.

If a set S is finite we denote by $\#S$ the number of elements in S.

Some Operations on Sets

Another method commonly used to define sets is to give a condition for membership in the set, so that the set consists of all objects in some fixed domain which satisfy the given condition. If S is the set so defined, and "$P(x)$" represents the statement "x satisfies the conditions," then we write:

$$S = \{x \,|\, P(x)\}$$

which is read, "S is the set of all x such that $P(x)$ is true." $P(x)$ may be given either in English or by means of logical formulas to be discussed later is this chapter.

Using this notation, we define the following operations on sets:

Union: $S \cup T = \{x \,|\, x \in S \quad \text{or} \quad x \in T \,(\text{or both})\}$.

Intersection: $S \cap T = \{x \,|\, x \in S \quad \text{and} \quad x \in T\}$.

Difference: $S \backslash T = \{x \,|\, x \in S \quad \text{and} \quad x \notin T\}$.

It is easily seen from the definitions that the following statements are true for any sets R, S, T:

1. $S \cup S = S$,
2. $S \cup T = T \cup S$,
3. $(R \cup S) \cup T = R \cup (S \cup T)$,
4. $S \cup \varnothing = S$,
5. $R \subseteq R \cup S$,
6. $R \cup (S \cap T) = (R \cup S) \cap (R \cup T)$,
7. $R \backslash (S \cup T) = (R \backslash S) \cap (R \backslash T)$,
8. $S \cap S = S$,
9. $S \cap T = T \cap S$,

10. $(R \cap S) \cap T = R \cap (S \cap T)$,
11. $S \cap \varnothing = \varnothing$,
12. $R \supseteq R \cap S$,
13. $R \cap (S \cup T) = (R \cap S) \cup (R \cap T)$,
14. $R \backslash (S \cap T) = (R \backslash S) \cup (R \backslash T)$,
15. If $R \subseteq S$ and $S \subseteq T$ then $R \subseteq T$,
16. $R \subseteq S$ if and only if $R \cup S = S$,
17. $R \subseteq S$ if and only if $R \cap S = R$.

Examples

(a) $\{0, 1, 2\} \cup \{-1, 0, 1, 2\} = \{-1, 0, 1, 2\}$,
$\{0, 1, 2\} \cap \{-1, 1, 2\} = \{1, 2\}$,
$\{1, 2\} \cap \{0, 3\} = \varnothing$,
$(0, 1, 2\} \backslash \{1, 3\} = \{0, 2\}$.

(b) $\{1, 2, 4\} \cup (\{1, 3, 5\} \cap \{2, 3, 6\}) = \{1, 2, 4\} \cup \{3\}$
$= \{1, 2, 3, 4\} = \{1, 2, 3, 4, 5\} \cap \{1, 2, 3, 4, 6\}$
$= (\{1, 2, 4\} \cup \{1, 3, 5\}) \cap (\{1, 2, 4\} \cup \{2, 3, 6\})$.

(c) $\{1, 2, 4\} \backslash (\{1, 3, 5\} \cup \{2, 3, 6\}) = \{1, 2, 4\} \backslash \{1, 2, 3, 5, 6\} = \{4\}$
$= \{2, 4\} \cap \{1, 4\} = (\{1, 2, 4\} \backslash \{1, 3, 5\}) \cap (\{1, 2, 4\} \backslash \{2, 3, 6\})$.

Exercises

1. Prove equalities 6 and 13.
2. Prove equalities 7 and 14.
3. Prove, or disprove by a counterexample:

(a) $(R \backslash S) \backslash T = R \backslash (S \backslash T)$,
(b) $(R \backslash S) \backslash T = R \backslash (S \cup T)$,
(c) $R \cup (S \cap T) = (R \cup S) \cap T$,
(d) $R \backslash S = R \backslash (R \cap S)$.

Cartesian Products

An **ordered pair** is a sequence (a, b) of two objects "a" and "b," not necessarily distinct. By definition, two pairs (a_1, b_1) and (a_2, b_2) are **equal** if and only if $a_1 = a_2$ and $b_1 = b_2$. If S and T are sets, their **cartesian product** is

$$S \times T = \{(a, b) \mid a \in S \quad \text{and} \quad b \in T\}.$$

Similarly, we speak of ordered triples, quadrupules, *n***tuples**, and so forth, as being sequences of the forms (a_1, a_2, a_3), (a_1, a_2, a_3, a_4), $(a_1, a_2, \ldots, a_n)$, and so forth. An *n*tuple $(a_1, a_2, \ldots, a_n)$ is defined to be equal to an *m*tuple $(b_1, b_2, \ldots, b_m)$ if and only if $m = n$ and $a_1 = b_1$, $a_2 = b_2, \ldots, a_n = b_m$. If $S_1, S_2, \ldots, S_n$ are all sets, their *n***fold cartesian product** is

$$S_1 \times S_2 \times \cdots \times S_n = \{(a_1, \ldots, a_n) \mid a_1 \in S_1, \ldots, a_n \in S_n\}.$$

If $S_1 = S_2 = \cdots = S_n = S$, this is written S^n.

For example, if $S_1 = \{a, b\}$, $S_2 = \{0, 1\}$, and $S_3 = \{?, !\}$, then

$$S_1 \times S_2 = \{(a, 0), (a, 1), (b, 0), (b, 1)\},$$
$$S_1 \times S_2 \times S_3 = \{(a, 0, ?), (a, 0, !), (a, 1, ?), (a, 1, !),$$
$$(b, 0, ?), (b, 0, !), (b, 1, ?), (b, 1, !)\},$$

and

$$S_2^2 = \{(0, 0), (0, 1), (1, 0), (1, 1)\}.$$

An Abbreviation

Since we will be working quite extensively with *n*tuples in this book, in order to shorten formulas we will use $\vec{x}_n$ to denote the sequence $x_1, x_2, \ldots, x_n$. Note that $\vec{x}_n$ does not include parentheses, so that the *n*tuple $(x_1, x_2, \ldots, x_n)$ is abbreviated to $(\vec{x}_n)$. Following this idea, $(\vec{x}_m, \vec{y}_n)$ denotes the $(m + n)$tuple $(x_1, x_2, \ldots, x_m, y_1, y_2, \ldots, y_n)$.

Exercises

1. Express $\#(S_1 \times S_2 \times \cdots \times S_n)$ in terms of $\#S_1, \ldots, \#S_n$ (assuming $S_1, S_2, \ldots, S_n$ are all finite).
2. Let S and T be sets. Give necessary and sufficient conditions for the following statements to be true: (a) $S \times T = \varnothing$; (b) $\#(S \times T) = \#S$; (c) $\#(S \cup T) = \#S + \#T$.

1.2 Functions

If A and B are any two sets, a **total function** from A **into** B is any correspondence f which associates with every element a of A a single **image** $f(a)$ in B. A **partial function** from A into B is a correspondence which associates a single image $f(a)$ in B with every element a of some subset A' of A. We call A' the **domain** of f, and say that $f(a)$ is **undefined** if $a \in A \setminus A'$. Clearly f is total if and only if f is partial with domain $A' = A$. Thus "total" is a special case of "partial."

If not otherwise specified, an arbitrary function is assumed to be total. The statement "f is a function from A into B" is abbreviated as "$f: A \to B$," read "f maps A into B."

Suppose $f: A \to B$ and $g: A \to B$ are partial functions. By definition, f and g are **equal** if and only if for each $a \in A$, either

(a)   both $f(a)$ and $g(a)$ are defined, and $f(a) = g(a)$; or
(b)   both $f(a)$ and $g(a)$ are undefined.

An *n***ary** function is any function f such that $f: A_1 \times A_2 \times \cdots A_n \to B$ for some sets $A_1, \ldots, A_n, B$. If $f(a_1, \ldots, a_n) = b$, we say that $a_1, \ldots, a_n$ are the **arguments** of f. We use the terms **unary**, **binary**, **ternary** as synonyms for 1ary, 2ary, and 3ary, respectively.

If $f: A \to B$, the **range** of f is $\{f(a) \mid a$ is in the domain of $f\}$, that is, the set of all values assumed by $f(a)$ as a assumes values from A.

A function $f: A \to B$ is said to be **onto** if the range of f is the entire set B, in other words, if for every $b \in B$ there is at least one element a of A such that $f(a) = b$. A function $f: A \to B$ is **one to one** (abbreviated as "1–1") if $f(a) = f(a')$ implies $a = a'$ for all $a, a' \in A$. In other words, no element of B is the image of two distinct elements of A.

If A and B are understood from context, it is common to refer to a function $f: A \to B$ as simply "the function $f(x)$," and similarly for functions which are binary, ternary, and so forth. This is slightly ambiguous, since $f(x)$ may refer to either the function as a whole, that is, the entire correspondence between A and B, or to the particular value in B which corresponds to the current value of x in A. Context should make it clear which meaning is intended.

Examples

Let $A = N \times N$ and $B = N$, where N is the set of nonnegative integers.

1. $f(x, y) = x + y$, $g(x, y) = x \cdot y$ are total binary functions.
2. $f(x, y) = x - y$ is a partial binary function with domain

$$D = \{(x, y) \mid x \geq y\}$$

(partial since N is not the set of all integers).

3. $f(x) = x - 3$ is a partial unary function.
4. $f(x, y) = x \div y$ is a partial binary function with domain

$$D = \{(x, y) | y \text{ is a divisor of } x\}.$$

5. $f(x) = 2x + 3$ is a total function, but is not onto.
6. If A is the class (that is, set) of all partial unary functions $f : N \to N$ from the nonnegative integers into themselves, a function $\alpha : A \times A \to A$ can be defined by

$$\alpha(f, g)(x) = \begin{cases} f(g(x)) & \text{if } g(x) \text{ and } f(g(x)) \text{ are both defined,} \\ \text{undefined} & \text{otherwise.} \end{cases}$$

This function is known as the composition operator, and $\alpha(f, g)$ is the **composition** of f with g. A common notation for $\alpha(f, g)$ is $f(g)$, or $fg(x)$.

Exercises

1. Find the range and domain of the following functions, whose arguments and values are nonnegative integers. Which are one to one? Which are onto?

    (a) $f(x) = 2x,$ (b) $f(x) = x - 5,$ (c) $f(x, y) = x^2 - y^2,$
    (d) $f(x, y) = x + (x + y)^2.$

2. Prove or disprove: The function $f(g(x))$ is total if and only if both f and g are total. Assume that $f, g : N \to N$.

2. ALPHABETS AND WORDS

Throughout these notes we are concerned with developing a theory of computability of functions and predicates whose variables and values range over all words on a finite alphabet. This includes, as a special case, functions and predicates on the nonnegative integers, since we can interpret words as integers in an effective manner.

A finite nonempty set is called an **alphabet**, and the elements of the set are called **symbols** (or **letters**). A **word** (or **string**) over an alphabet A is any finite, possibly empty, sequence of symbols from A, written without punctuation. For example, if $A = \{0, 1\}$, then the following are words over A : 11, 0001, 101010, 0. The **empty word** is the empty sequence of symbols, and is written λ. The **length** of a word x (written $|x|$) is the number of symbols in x, including repetitions. Thus the words 11, 0001, 10101, 0, λ have lengths of 2, 4, 5, 1, 0, respectively. The set of all words over A is denoted by A^*. If $A = \{0, 1\}$, then $A^* = \{\lambda, 0, 1, 00, 01, 10, 11, 000, 001, \ldots\}$.

If $x = a_1 a_2 \cdots a_m$ and $y = b_1 \cdots b_n$ are words over A, and each a_i and b_j is in A, then the **catenation** of x and y is the word $a_1 a_2 \cdots a_m b_1 \cdots b_n$, denoted by xy. Throughout these notes, xy will denote catenation and not multiplication. The following are immediate consequences of these definitions, for any $x, y, z \in A^*$:

    (i) $x(yz) = (xy)z$,

    (ii) $x\lambda = x = \lambda x$,

    (iii) $|xy| = |x| + |y|$,

    (iv) $xy \neq yx$ in general, unless $\sharp A = 1$.

As with multiplication, we use a power notation for repeated catenation, so $x^3 = xxx$ and $(xy)^2 = xyxy$ for all $x, y \in A^*$. By definition, $x^0 = \lambda$ for any $x \in A^*$. From this we see that $|x^n| = n|x|$ and that $x^{m+n} = x^m x^n$ for any $x \in A^*$ and any nonnegative integers m, n. In general α^n will denote catenation if α is a string, but a set of $n$tuples if α is a set.

Number Notation

While we shall be concerned primarily with functions and predicates on words, on occasion it will be convenient to interpret words in terms of nonnegative integers. This will be done by means of the *m*adic number notation, defined as follows. This is called **dyadic** notation if $m = 2$.

Let A be an alphabet containing m symbols, where $m \geq 1$. Whenever words over A are to be interpreted numerically we shall assume that the symbols of A have been given a fixed ordering $a_1, a_2, \ldots, a_m$; further, each symbol a_i of A will be considered as a single digit, representing the integer i. Thus we may write $A = \{1, 2, \ldots, m\}$.

The function $\alpha : A^* \to N$, such that $\alpha(x)$ denotes the **numeric interpretation** of a word $x \in A^*$, is defined as follows:

(a) $\alpha(\lambda) = 0$, so the empty word is interpreted as zero; and
(b) if $x = i_k i_{k-1} \cdots i_1 i_0$, where each $i_j \in A$, then

$$\alpha(x) = i_0 + i_1 \cdot m + i_2 \cdot m^2 + \cdots + i_k \cdot m^k = \sum_{j=0}^{k} i_j \cdot m^j.$$

This representation of numbers is known as the *m*adic number notation. It has the desirable property that it is a one-to-one correspondence between A^* and N, so that every word over A is assigned a unique integer in N, and every integer in N is $\alpha(x)$ for exactly one word x over A. This is not true of conventional decimal or binary notations, since the words 13, 013, and 0013 all denote the same integer.

For an example, let $A = \{1, 2\}$. Then $\alpha(1) = 1$, $\alpha(2) = 2$, $\alpha(121) = 1 \cdot 4 + 2 \cdot 2 + 1 = 9$, and the words corresponding to the sequence 0, 1, 2, 3, ... are: λ, 1, 2, 11, 12, 21, 22, 111, 112, ...

Exercises

1. Under what conditions is it true that $(xy)^n = x^n y^n$?
2. Express $\alpha(xy)$ in terms of $\alpha(x)$, $\alpha(y)$, and $|y|$ (remember, xy denotes catenation).
3. Prove that $\alpha(x) = \alpha(y)$ if and only if $x = y$.
4. Prove that α is onto; that is, for every $n \in N$ there is a word $x \in A^*$ for which $\alpha(x) = n$.
5. Exercises 3 and 4 imply that there is a unique function $\beta : N \to A^*$ such that for all x, $\alpha(x) = n$ if and only if $\beta(n) = x$. Devise a procedure which will calculate $\beta(n)$, if given n. Assume $A = \{1, 2\}$ for simplicity.

3. PREDICATES

In this section we will suppose that there is a fixed, prespecified set U, such that any variable x which we discuss takes on values from U.

An *n*ary **predicate** $P(x_1, \ldots, x_n)$ is a relation involving n ordered variables $x_1, \ldots, x_n$, which is either *true* or *false* for any particular choice of values for $x_1, \ldots, x_n$ from U. If $U = N$, one example of a predicate is the unary relation $P(x)$ which is true if and only if x is a prime number. The relation $x < y$ is a binary predicate over U, and the equation $x^n + y^n = z^n$ can be viewed as a 4ary predicate $P(x, y, z, n)$. If $x = 12$, $y = 5$, $z = 13$, and $n = 2$, the values of these predicates become false, false, and true, respectively.

Any set $S \subseteq U$ may be considered as a unary predicate by interpreting

$$S(x) \text{ is true} \quad \text{if} \quad x \in S;$$

and

$$S(x) \text{ is false} \quad \text{if} \quad x \in U \backslash S.$$

Note: Students who do not plan to study the formal, detailed arguments of Chapters III, IV, V, and VI may omit the remainder of this section, as predicates are not essential to the informal arguments and overview of Chapter II. That material will be correct (and somewhat simpler) if interpreted in terms of sets, that is, unary predicates.

Boolean Operations

We now define various operations on predicates which make it possible to define complex predicates in terms of simpler predicates. The *Boolean* operations are shown in Table I.1. Their meanings are

TABLE I.1

Name	Symbol	English Reading
negation	$\neg$	"not"
conjunction	$\wedge$	"and"
disjunction	$\vee$	"or"
implication	$\Rightarrow$	"implies," or "if ..., then ..."
equivalence	$\Leftrightarrow$	"equivalent," or "if and only if"

defined as follows: If $P(\vec{x}_n)$ and $Q(\vec{x}_n)$ are $n$ary predicates, then $\neg P(\vec{x}_n)$ is true when $P(\vec{x}_n)$ is false, and false when $P(\vec{x}_n)$ is true. $P(\vec{x}_n) \wedge Q(\vec{x}_n)$ is true if both $P(\vec{x}_n)$ and $Q(\vec{x}_n)$ are true, and is false otherwise. $P(\vec{x}_n) \vee Q(\vec{x}_n)$ is true if either $P(\vec{x}_n)$ is true or $Q(\vec{x}_n)$ is true, or both; and false if both are false. $P(\vec{x}_n) \Rightarrow Q(\vec{x}_n)$ is false only when $P(\vec{x}_n)$ is true and $Q(\vec{x}_n)$ is false; otherwise it is true. $P(\vec{x}_n) \Leftrightarrow Q(\vec{x}_n)$ is true if $P(\vec{x}_n)$ and $Q(\vec{x}_n)$ are either both true or both false, and otherwise false.

Note: Throughout these notes we shall often use "iff" as an informal abbreviation for the phrase "if and only if."

Explicit Transformation

The operations other than $\neg$ require that P and Q have the same number of variables in the same order. We now introduce another operation, called *explicit transformation*, which allows variables to be permuted, identified, or replaced by constants, and allows dummy variables to be added.

We say that $P(\vec{x}_n)$ is an **explicit transformation** of $Q(\vec{y}_m)$ if there are $\xi_1, \xi_2, \ldots, \xi_m$, each of which is either an element of U or one of the variable names $x_1, x_2, \ldots, x_n$, such that for all values $x_1, x_2, \ldots, x_n \in U$, $P(\vec{x}_n)$ is true if and only if $Q(\vec{\xi}_m)$ is true.

Following are several examples of explicit transforms of a ternary predicate $Q(y_1, y_2, y_3)$.

(i) $P1(x_1, x_2, x_3)$ iff $Q(x_2, x_3, x_1)$ (permutation of variables)

(ii) $P2(x_1, x_2)$ iff $Q(x_1, x_2, x_1)$ (identification of variables)

(iii) $P3(x_1, x_3)$ iff $Q(x_1, 13, x_3)$ (replacement of a variable by a constant)

(iv) $P4(x_1, x_2, x_3, x_4)$ iff $Q(x_1, x_2, x_3)$ (addition of a dummy variable)

(v) $P5(x_1, x_2, x_3)$ iff $Q(x_1, 13, x_1)$ (combination of the above).

Suppose that an "oracle" for Q is given, that is, some device which, when given any m-tuple $(\vec{y}_m)$, will answer "true" if $Q(\vec{y}_m)$ is true, and "false" if not. If P is an explicit transform of Q, as above, then the oracle for Q can be used for P as follows:

(a) Given $x_1, \ldots, x_n$ in U, first construct the m-tuple $(\xi_1, \ldots, \xi_m)$.
(b) Apply the oracle to $(\xi_1, \ldots, \xi_m)$.
(c) Answer "$P(\vec{x}_n)$ is true" if the oracle's answer is "true"; otherwise answer "$P(\vec{x}_n)$ is false."

Quantifiers

Let $P(\vec{x}_n, y)$ be an $(n + 1)$ary predicate. The **existential quantification** of P is an $n$ary predicate denoted by $\exists y \, P(\vec{x}_n, y)$, whose variables are $x_1, \ldots, x_n$. For any choice of $x_1, \ldots, x_n$ from U, $\exists y \, P(\vec{x}_n, y)$ is true iff there is at least one y in U for which $P(\vec{x}_n, y)$ is true. The **universal quantification** of P is an $n$ary predicate denoted by $\forall y \, P(\vec{x}_n, y)$. For any choice of $x_1, \ldots, x_n$ from U, $\forall y \, P(x_n, y)$ is true iff $P(\vec{x}_n, y)$ is true for every $y \in U$. In both cases y is merely a dummy, called a **bound variable**. In fact, the formulas $\exists y \, P(\vec{x}_n, y)$ and $\exists z \, P(\vec{x}_n, z)$ denote the same predicate.

We read $\exists y \, P(\vec{x}_n, y)$ as "there exists a y such that $P(\vec{x}_n, y)$," and $\forall y \, P(\vec{x}_n, y)$ as "for all y, $P(\vec{x}_n, y)$."

Formulas

By analogy with the formulas of algebra which involve arithmetic operations, $+, -, \cdot, \div$, and define functions, we may define predicates by formulas involving the logical operations just mentioned (except for explicit transformation, which is used implicitly). The rules for constructing formulas over an alphabet A are the following:

(i) If P is a symbol representing an $n$ary predicate, and each of $t_1, t_2, \ldots, t_n$ is either a variable name or a word over A, then $P(t_1, \ldots, t_n)$ is a formula.
(ii) If φ is a formula, then $(\neg \varphi)$, $(\exists y \, \varphi)$, and $(\forall y \, \varphi)$ are also formulas.
(iii) If φ and ψ are formulas, then $(\varphi \wedge \psi)$, $(\varphi \vee \psi)$, $(\varphi \Rightarrow \psi)$, and $(\varphi \Leftrightarrow \psi)$ are also formulas.

These rules require that every subformula of a given formula must be enclosed in parentheses, in order to make the grouping unambiguous. In general, parentheses may be omitted, subject to the following interpretations. Within a formula F, define the **scope** of an occurrence of a logical operator to consist of that portion of F to which the operator applies. For example, within the algebraic formula $a + b * (c/d)$, the scope of " $/$ " is c/d, the scope of " $*$ " is $b * (c/d)$, and the scope of " $+$ " is the entire formula.

Using this concept, the following rules may be used for interpretation of a logical formula which is not fully parenthesized.

1. The scope of $\neg$, $\forall y$, and $\exists y$ is the *smallest* subformula which occurs immediately to the right of these symbols.
2. The operators $\wedge$, $\vee$, $\Rightarrow$, and $\Leftrightarrow$ have successively larger scopes, analogous to those of the arithmetic operators $\cdot$ and $+$ (for example, if $\wedge$ and $\vee$ occur near one another without intervening parentheses, the scope of the $\vee$ includes the scope of the $\wedge$).
3. A series of occurrences of the same operator is grouped from left to right [for example, $A \Rightarrow B \Rightarrow C$ is read $(A \Rightarrow B) \Rightarrow C$].

Examples of Formulas

In the following, " $x < y$ " denotes the binary predicate over N which is true iff x is less than y; and similarly for " $x \leq y$," and so forth.

$P(x) \wedge Q(x, z, y)$;

$x = y \vee x < z \wedge y < z$ [equivalent to $x = y \vee (x < z \wedge y < z)$];

$x = y \wedge x < z \vee y < z$ [equivalent to $(x = y \wedge x < z) \vee y < z)$];

$x \leq y \Rightarrow x \neq y \Rightarrow x < y$ [equivalent to $(x \leq y \Rightarrow x \neq y) \Rightarrow x < y$];

$\neg P(x) \vee \exists y \neg x < y \wedge x = 0$

           [equivalent to $(\neg P(x)) \vee (\exists y(\neg (x < y)) \wedge x = 0)$].

It is convenient to define new predicates from previously defined predicates by formulas using the equivalence operation $\Leftrightarrow$. For instance, let $U = N$ and suppose the predicate $u < v$ has already been defined.

We can define the predicate $B(x, y, z)$ (y *between* x and z) from $u < v$ as follows:

$$B(x, y, z) \Leftrightarrow (x < y \wedge y < z) \vee (z < y \wedge y < x).$$

The truth or falsity of $B(1, 3, 2)$ may be determined by simply substituting 1 for x, 3 for y, and 2 for z in the right side of the formula above.

It should be remarked that this use of formulas hides many applications of the explicit transformation operations. For example, to show that B can be constructed from explicit transformation, $\wedge$, $\vee$, and the relation $u < v$ as above, the following explicit sequence would be necessary:

1. $P_1(x, y, z)$ iff $x < y$ (explicit transformation of " $<$ "),
2. $P_2(x, y, z)$ iff $y < z$ (explicit transformation of " $<$ "),
3. $P_3(x, y, z)$ iff $P_1(x, y, z) \wedge P_2(x, y, z)$ (conjunction),
4. $P_4(x, y, z)$ iff $z < y$ (explicit transformation of " $<$ "),
5. $P_5(x, y, z)$ iff $y < x$ (explicit transformation of " $<$ "),
6. $P_6(x, y, z)$ iff $P_4(x, y, z) \wedge P_5(x, y, z)$ (conjunction),
7. $B(x, y, z)$ iff $P_3(x, y, z) \vee P_6(x, y, z)$ (disjunction).

Exercises

1. Express $\Leftrightarrow$ and $\Rightarrow$ in terms of $\vee$, $\wedge$, and $\neg$.
2. Express $\forall$ in terms of $\exists$ and $\neg$.
3. Express $\wedge$ in terms of $\vee$ and $\neg$.
4. Prove each of the following formulas either true or false (give a counterexample if false). Let $U = N$.
    (a) $\forall x \, \forall y \, \exists z (x + y = z)$,
    (b) $\forall x \, \forall y \, \exists z (x + z = y)$,
    (c) $\exists x \, P(x) \vee \exists y \, Q(y) \Leftrightarrow \exists z (P(z) \vee Q(z))$,
    (d) $\forall x \, P(x) \vee \forall y \, Q(y) \Leftrightarrow \forall z (P(z) \vee Q(z))$,
    (e) $\forall x \, P(x) \wedge \forall y \, Q(y) \Leftrightarrow \forall z (P(z) \wedge Q(z))$.
5. Construct formulas which define each of the following predicates in terms of: $x = 0$, $x + y = z$, and $x \cdot y = z$. Assume $U = N$.
    (a) x is even,
    (b) $x \leq y$,

(c) $x < y$,

(d) x is a divisor of y,

(e) x is a power of 2,

(f) x is prime,

(g) the greatest common divisor of x and y is z,

(h) z is the remainder on division of x by y.

Function Definition by Minimalization

A unary predicate $P(x)$ over A corresponds naturally to the set $\{x \,|\, P(x)$ is true$\}$. In an analogous way we may associate a *function* with each $(n + 1)$ary predicate, as follows.

Let $Q(\vec{x}_n, y)$ be a predicate, and interpret words over $A = \{1, 2, \ldots, m\}$ as $m$adic integers as in Section I.2. A function $f(\vec{x}_n)$ is obtained from Q by **minimalization**, or the μ-operator iff for every $x_1, \ldots, x_n \in A^*$

$$f(\vec{x}_n) = \begin{cases} \text{the least } y \text{ such that } Q(\vec{x}_n, y) \text{ is true if such a } y \text{ exists,} \\ \text{undefined} \quad \text{if} \quad Q(\vec{x}_n, y) \text{ is false for all } y. \end{cases}$$

This is written as

$$f(\vec{x}_n) = \mu y Q(\vec{x}_n, y)$$

For example, let $Q(x, y, z)$ denote the predicate $x \cdot y = z$. Then the function $f(x, y) = \mu z Q(z, y, x)$ will equal $x \div y$ if x is divisible by y, and will be undefined otherwise.

Exercise

Construct formulas which define the following functions, using minimalization and the predicates $x + y = z$ and $x \cdot y = z$ (assume $U = N$).

(a) $f(x, y) = \begin{cases} x - y & \text{if} \quad x \geq y, \\ \text{undefined} & \text{otherwise.} \end{cases}$

(b) $g(x) = \begin{cases} \sqrt{x} & \text{if} \quad x \text{ is a perfect square,} \\ \text{undefined} & \text{otherwise.} \end{cases}$

4. INDUCTION AND INDUCTIVE DEFINITION

The concept of proof by induction and the idea of inductive definition of sets, classes of functions, classes of predicates, and so forth are fundamental to the theory of computability. For instance, in these notes the classes of S-rudimentary and existentially definable predicates, and the μ-recursive functions will be defined inductively, and will be shown recursive or recursively enumerable by inductive proofs.

More generally, the aim of these notes is to study the class of all *effective processes*, that is all processes which are in some sense mechanically executable by purely finitary means. For example, all functions which can be computed by an idealized general-purpose digital computer would be effectively computable in this sense. Such a finitary process can be broken down into a (perhaps very long) series of very elementary operations. We wish to study the computing power or expressiveness of long series of elementary actions of various types. This can be done by use of an inductive definition of computability, such as the following for functions:

   (i)   Each elementary function is computable (either because it is very simple, or by definition).

   (ii) If a function f is computable, and if the function g can be obtained by applying an elementary operation to f, then g is also computable.

   (iii) A function is computable only if it can be obtained from (i) and (ii).

Similar schemata can be constructed to define the computability of predicates or more general processes.

4.1   Mathematical Induction

Suppose that $P(k)$ is some proposition which depends only upon a nonnegative integer k, and that for each value of k in N, $P(k)$ is either true or false. The **principle of mathematical induction** is that $P(0)$, $P(1)$, ... are all true, if the following two conditions are satisfied:

(a) $P(0)$ is true, and
(b)   if we let k denote an arbitrary number in N, and if we assume that $P(k)$ is true, then it follows logically that $P(k + 1)$ must also be true.

Example

Let A be the set of all ordinary algebraic expressions which can be constructed from real numbers and a single variable x, by use of the operations $+$ and $\cdot$ (multiplication). Define a *polynomial* to be any expression of the form $a_0 + a_1 x + a_2 x^2 + \cdots + a_n x^n$, where $a_0, \ldots, a_n$ are real coefficients, and say that two expressions $e_1(x)$ and $e_2(x)$ are *equal* iff $e_1(a) = e_2(a)$ for every real number a.

PROPOSITION

Any expression in A is equal to a polynomial.

PROOF

Define the **height** of an expression to be the number of occurrences of $+$ and $\cdot$ in it. Clearly, every expression in A has a unique nonnegative height. For example, the heights of $(x + 5)$, $(10 + x) \cdot (x + 5)$, and 137 are 1, 3, and 0, respectively.

Now define the *inductive hypothesis* IH(k) by: IH(k) is true if every expression which has height not exceeding k is equal to a polynomial.

1.   IH(0) is true, since if $e(x)$ has height not exceeding 0, then either $e(x) = x$ or $e(x) = a$ for some real a. In either case $e(x)$ is a polynomial itself.
2.   Suppose IH(k) is true. *To show*: IH($k + 1$) must also be true. Let expression $e(x)$ have height $k + 1$. Then, by definition of A, there must exist expressions $p(x)$ and $q(x)$ of height less than $k + 1$ such that either $e(x) = p(x) + q(x)$, or $e(x) = p(x) \cdot q(x)$.

By IH(k), there are polynomials such that $p(x) = a_0 + a_1 x + \cdots + a_m x^m$ and $q(x) = b_0 + \cdots + b_n x^n$. In the first case, let

$$c_i = \begin{cases} a_i + b_i & \text{if } 1 \le i \le \min(n, m), \\ a_i & \text{if } n < i \le l, \\ b_i & \text{if } m < i \le l, \text{ where } l = \max(m, n). \end{cases}$$

Then $e(x) = c_0 + c_1 x + \cdots + c_l x^l$. In the second case, it is easily seen that

$$e(x) = \sum_{k=0}^{m+n} \left(\sum_{i+j=k} a_i b_j \right) x^k,$$

so in either case $e(x)$ is equal to a polynomial. Thus IH($k + 1$) follows from IH(k). Thus IH(k) is true for all k. ∎

Exercises

1. Prove that, for $n = 0$, 1, 2, ..., $1^3 + 2^3 + 3^3 + \cdots + n^3 = n^2(n + 1)^2/4$.
2. Prove that any expression constructed as above with operators $+$, $\cdot$, and $/$ ("divide") is equal to the quotient of two polynomials.
3. Prove that any formula (as defined in Section 3) is equivalent to a formula which involves only the operators $\vee$, $\neg$, and $\exists$. *Hint:* Use Exercises 1, 2, and 3 of Section 3.

4.2   Inductive Definitions

Consider the following definition of a set T of words over the alphabet $A = \{x, +, \cdot, (,), 1, 2, 3\}$:

1. x, 1, 2, and 3 are in T.
2. If u and v are words in T, then so are $(u + v)$ and $(u \cdot v)$.
3. No words are in T unless they can be shown to be in T by rules 1 and 2.

For example, x and 3 are in T by rule 1, and so $(x + 3)$ and $((x + 3) \cdot x)$ are in T by rule 2. Clearly, T consists of all expressions involving x, 1, 2, and 3, and the operators $+$ and $\cdot$, subject to the condition that for each operator there is a pair of parentheses delimiting its scope.

In general, an inductive definition is given in three parts. Assume that all objects are in some fixed universal set U.

(i) One or more **basis clauses**, each naming certain objects which are in T by definition.

(ii) Some number of **inductive clauses**, usually of the form: If $x_1, \ldots, x_n$ are in T, then so is $f(x_1, \ldots, x_n)$, where $f: U^n \to U$ is a given (possibly partial) $n$ary function (for example, the case above involves two functions on $U = A^*$, namely $f(u, v) = (u + v)$, and $g(u, v) = (u \cdot v)$. Note that f, g have *words* as values, *not* numbers.)

(iii) An **extremal** clause, which states that T contains only objects which can be shown to be in T by some number of applications of the basis and inductive clauses. This clause is often omitted.

Let $f: U^n \to U$ be a partial function. We say that T is **closed under** f if and only if $f(\vec{x}_n)$ is in T whenever $x_1, \ldots, x_n \in T$ and $f(\vec{x}_n)$ is defined. It is easily seen that T is the smallest set which contains the basis objects and is closed under the operations of its inductive clauses. We now give more examples.

1. Let $U = N$ (the set of nonnegative integers), and define T by

    (i) $0 \in T$ and $1 \in T$.
    (ii) $x \in T$ and $y \in T$ implies that $x + y \in T$ (now $x + y$ denotes the *value*, or sum of x and y, since $U = N$).
    (iii) Extremal clause.

    Then $T = N$.

2. Replace (i) by $4 \in T$. Then T consists of all positive multiples of 4.

3. Replace (i) by $6 \in T$ and $8 \in T$. Then $T = \{6, 8, 12, 14, 16, 18, 20, \ldots\} = \{x \mid x \text{ is even and } x > 4 \text{ and } x \neq 10\}$.

4. Let $U = \{f \mid f: N \to N\}$, and define T by

    (i)   The functions $e(x) = x$ and $S(x) = x + 1$ are in T.

(ii)   If f and g are in T, then so is h, where $h(x) = f(g(x))$
for all $x \in N$. *Note:* $h(x)$ denotes a function in one variable
x, not the value of h as applied to a specific value of x.

(iii) Extremal clause.

In this case T consists of all functions of the form $f(x) = x + a$,
where $a \in N$. To prove this, observe (by induction on a) that
$e(x) = x + 0$ is in T, and that if $f(x) = x + a$ is in T, then $S(f(x))$
$= (x + a) + 1 = x + (a + 1)$ is in T. Thus T contains all functions $f(x) = x + a$. For the converse, note first that the basis
functions are of the required form. Second, if $f(x) = x + a$ and
$g(x) = x + b$, then $f(g(x)) = x + (a + b)$, and so that any function in T is of the form $f(x) = x + a$.

Exercises

1.   State in explicit form the inductive hypothesis used in the
proof which was just given.

2.   Find a general form for the set T of functions defined by

(i) $0(x) = 0$, $e(x) = x$, $S(x) = x + 1$ are in T;
(ii)   if f, g are in T then so is $h(x) = f(g(x))$ and $j(x) = f(x)$
$+ g(x)$;
(iii) extremal clause.

Prove that T consists of all and only functions which can be
expressed in your general form.

3.   Give an inductive definition of the set of all functions of the
forms $f(x) = ax + b$ or $g(x, y) = ax + by + c$, where $a, b, c \in N$.

5. COUNTABILITY AND ENUMERATION FUNCTIONS

5.1   Sets

DEFINITION I.5.1

A set S is **countable** iff S is empty, or if there is a sequence s_0, s_1, s_2,
s_3, ... which contains *all* and *only* the elements of S.   ∎

TABLE I.2 EXAMPLES

Set	Corresponding Enumeration Sequence
$N = \{0, 1, 2, \ldots\}$	$0, 1, 2, 3, \ldots$
$Z = \{\ldots -2, -1, 0, 1, 2, \ldots\}$	$0, 1, -1, 2, -2, 3, -3, \ldots$
A^*, for $A = \{1, 2\}$	$\lambda, 1, 2, 11, 12, 21, 22,$
	$111, 112, 121, 122, 211, 212, 221, 222, \ldots$
$N \times N = \{(x, y): x, y \in N\}$	$(0, 0), (0, 1), (1, 0), (0, 2), (1, 1), (2, 0),$
	$(0, 3), (1, 2), (2, 1), (3, 0), \ldots$

REMARKS

1. An alternate (and equivalent) definition is that S is countable iff it is finite, or can be placed in a one-to-one correspondence with the natural numbers.

2. The sequence $s_0, s_1, s_2, \ldots$ may contain the same element more than one time (for example the sequence $0, 1, 2, 2, 3, 3, 3, \ldots$ lists all and only the elements of N). It should be clear that a new sequence could be constructed by removing all repeated elements, so that if S is infinite and countable, it is also countable without repetitions.

3. It is possible that the elements of S can be arranged in a sequence $s_0, s_1, s_2, \ldots$ *in principle*, even though there is no effective, algorithmic way to do this *in practice*. Such sets will be discussed in Section II.1 and Chapter V.

The sequence $s_0, s_1, s_2, \ldots$ may be considered as the range of the function f, defined by: $f(0) = s_0, f(1) = s_1, \ldots$. Such a function is said to *enumerate* S.

DEFINITION I.5.2

Let $S \subseteq N$, and let $f: N \to N$ be a function. We say that f **enumerates** S iff $S = \{f(0), f(1), f(2), \ldots\}$. ∎

Clearly S is countable iff $S = \emptyset$ or S has an enumerating function.

Exercises

1. Show that the set of rational numbers is countable. Note: x is rational iff $x = p/q$ for some integer p and positive integer q.

2. Show that the following set is countable:

    $$S = \{(c_1, c_2, \ldots, c_n) \mid n > 0 \quad \text{and} \quad c_1, \ldots, c_n \in N\}.$$

    Typical elements of S are

    $$(1, 2), (17), (0, 0, 0, 0), (3, 2, 1776).$$

3. A real number is called *algebraic* iff it is a root of a polynomial in one variable which has integer coefficients. Show that the collection of all algebraic numbers is countable. *Hint:* Use the result of Exercise 2.
4. Show that if S and T are countable sets, then $S \cup T$ and $S \times T$ are also countable.
5. Construct a unary function $f(n)$ which enumerates the set of pairs (x, y) such that $x, y \in N$. Give an explicit definition of f, rather than just an enumeration of its range.

An Uncountable Set

The proof of the following theorem uses a technique called **diagonalization** which is central to much of computability theory.

THEOREM I.5.3

The set of all subsets of $N = \{0, 1, 2, \ldots\}$ is not countable.

PROOF

Let $\mathbf{N}$ denote the set of all subsets of N. Suppose, for the sake of argument, that $\mathbf{N}$ is countable, so there is a sequence $N_0, N_1, N_2, \ldots$ which includes all and only the subsets of N.

Construct a new set M as follows:

$$M = \{n \mid n \notin N_n \quad (n = 0, 1, 2, \ldots)\}.$$

Thus M contains 0 only if 0 is not in N_0, 1 only if 1 is not in N_1, and so forth. Clearly M is a subset of N, so by the assumption that $N_0, N_1, N_2, \ldots$ enumerated *all* subsets of N, we must have $M = N_k$ for some value k.

But this is impossible, since the natural number k is in M if and only if k is not in $N_k = M$ (by definition of M). Thus the assumption that $\mathbf{N}$ could be enumerated leads to a contradiction, so we must conclude that $\mathbf{N}$ is not countable.   ∎

Exercise

Show the set of all real numbers between 0 and 1 is not countable.

Hint: Consider their binary representations, and assume that these could be enumerated. Construct a real number which cannot be in the enumeration, by a technique similar to the above.

5.2   Functions

It we apply the ideas of the previous section to sets of functions, some minor difficulties arise because of notation and partially defined functions. To circumvent this we introduce the following:

DEFINITION I.5.4

Let A, B be sets and let S be a collection of partial functions $f : A \rightarrow B$. We say that a binary function $u : N \times A \rightarrow B$ **enumerates** S iff whenever $g : A \rightarrow B$ is an arbitrary function then g is in S *if* and *only if* there is a natural number k such that

$$g(x) = u(k, x) \qquad \text{for all} \quad x \in A.$$

Further, we say that S is **countable** iff S has an enumerating function, or S is empty. If S is countable then $u(0, x)$, $u(1, x)$, $u(2, x)$, ... is a sequence which contains all and only the functions in S.   ∎

Clearly u must be a total function if every function in S is total.

Examples

1. S consists of all functions of the forms $2x + 3$, $4x + 3$, $6x + 3$, and so forth. An enumerating function is given by

$$u(k, x) = (2k + 2) \cdot x + 3.$$

2. S consists of the functions $0x - 0$, $x - 1$, $2x - 2$, $3x - 3$, and so forth. Note that all of these except the first are partial on N. An enumerating function is $u(k, x) = k \cdot x - k$.

THEOREM I.5.5

The set of all unary total functions from N to N is uncountable.

PROOF

Suppose that S is countable, with enumerating function $u(k, x)$. Define

$$g(x) = u(x, x) + 1 \qquad \text{for all} \quad x \in N.$$

Clearly g is a total unary function, so that g must be in S. This means that for some k,

$$g(x) = u(k, x) \qquad \text{for all} \quad x \in N.$$

But this is clearly impossible, for if we set x to k we obtain

$$u(k, k) + 1 = g(k) = u(k, k);$$

that is,

$$1 = 0.$$

Thus the assumption that S is countable must be false.   ▌

This is another example of proof by diagonalization. Note that the argument above would fail if S were taken to be the set of all unary *partial* functions, since the equation $u(k, k) + 1 = u(k, k)$ is not a contradiction if $u(k, k)$ is undefined. However, the proof above shows that the set of all unary partial functions contains an uncountable subset, and so it is itself uncountable (recall that every total function is also partial).

Exercises

1. Construct an enumerating function for the set containing the functions: $x + 1, 4x + 3, 9x + 5, 16x + 7, \ldots$.
2. Construct an enumerating function $f(k, x)$ the collection of all functions of the form $ax + b$, for $a, b \in N$. *Hint:* Use the result of Exercise 5, Subsection I.5.1.
3. Consider the set of functions of Exercise 1. If the construction of Theorem I.5.5 is applied to its enumerating function, what function g do you obtain?

■ INTRODUCTION TO COMPUTABILITY

Our aim is the study of the class of all *effective processes*, that is, all processes which can be performed in a determinate, precisely specified manner, using only steps which are mechanically executable by finitary methods. In particular we want to define the class of all effectively computable functions and the class of all effectively decidable predicates, and to establish the basic properties and limits of these classes.

Such a statement of purpose immediately arouses many questions, such as the following:

(a)   Isn't the purpose too vaguely stated to be treated meaningfully?
(b)   Just what do we mean by "computable," or "effective"?
(c)   Could any single definition possibly be adequate? Or would any definition be too narrow because it would exclude computing devices with special capabilities?

These questions are quite important, and need to be answered. To do this, we shall give a single definition of effectiveness, in terms of the Turing machine. All above questions would be answered if we could prove, somehow, that every intuitively effective process can be carried out by a Turing machine, and that the Turing machine's computations are effective.

In one direction this is easy, since we will see that the Turing machine is sufficiently simple that its computations are certainly effectively computable in any reasonable sense. Unfortunately, we cannot prove formally that every intuitively effective process is Turing computable, because we have no clear-cut, precise definition of an intuitively effective process. However, we will present some rather convincing mathematical evidence of several types that the two concepts are identical. These arguments for the generality of Turing machine computability will be stated at the end of this chapter, and will form a large part of the justification for the generality claimed for the material in later chapters.

1. IMPLICATIONS OF THE CONCEPT OF EFFECTIVENESS

Just what is meant by the term "effective process"? In a formal context we shall later characterize the concept by means of the Turing machine. In this section we discuss the concepts of effectiveness and effective computability, enumerability, and decidability, all in an informal, intuitive way. We shall show that various interesting consequences follow logically from the natural informal definitions of these terms.

The arguments used to establish these consequences embody the reasoning which is used (in a more formal context) to prove the fundamental results of computability theory, and so will be an intuitive preview of many of the theorems of Chapter V. These results will be called "propositions" to distinguish them from their formal analogs which occur later.

1.1 Effective Processes and Effectively Computable Functions

First, an "effective process" is a general problem-solving method, which is used to solve any one of a class of questions to which it is applicable. The term is not applied to a single question, such as "what is the minimum number of colors needed to unambiguously color the regions of a planar map?" for there is only a single answer to this question (even though we do not know yet whether this answer is 4 or 5).

Further, the process must give the correct answer to each question which is within its domain of applicability. Its behavior when applied to questions outside this domain is not of interest, except that the process is not allowed to give incorrect answers. However, it may fail to terminate, or may terminate in a nonstandard way.

Second, the process must be one which can be performed in a purely finitary manner, one step at a time. It must be **deterministic**, so that at each step there is at most a single next step to perform (that is, no "guessing" is required).

It cannot require the completion of an infinite number of steps, or a continuous series of steps. In addition, each step must itself be purely finitary, so a single step cannot depend upon an infinite amount of data for its completion, or the completion of an infinite process. An effective process could, in principle, be performed on a purely mechanical basis. That means that a machine (for example, a digital computer) could be constructed or programmed or instructed in advance to perform the process without need for human decision-making once the machine has begun operation.

The process is required to terminate eventually if applied to any data which is within its domain of applicability.

Finally, the process must possess a description which is finite in size, and can be executed in a mechanical, finitary way. For example an infinite list of instructions to a machine would not be considered to specify an effective process unless it could be replaced by an equivalent finite list.

Note that we do not set any *a priori* bound on the complexity of the steps themselves. We only require that the sequence be finite and finitely describable, and that the steps be mechanically executable.

We informally define an **algorithm** to be a finite executable description of an effective process. An algorithm can be expressed in an appropriate language, such as English, mathematical symbols, or an artificial language used to program computers, such as Fortran or Algol ("Formula Translator" and "Algorithmic Language," respectively). A familar example of an algorithm is a recipe for making cookies. The remarks above imply that any effective process is described by some algorithm.

A partial or total *n*ary function *f* is said to be **effectively computable** if there is an effective process which, when given any *n* argument values $x_1, \ldots, x_n$, will either

1. eventually halt, yielding $f(\vec{x}_n)$ if $f(\vec{x}_n)$ is defined, or
2. never halt if $f(\vec{x}_n)$ is undefined.

For example, the function $f(x, y) = x + y$ (where x, y, and $x + y$ are expressed in decimal notation) is effectively computable by the algorithm for digit-by-digit addition which is given in elementary schools.

It might seem that the definition of an effectively computable function depends on the notation used to represent the arguments (for example, decimal, binary, prime power representation). However, this makes no difference as long as there is an effective way of translating from one notation to another and back. For example, suppose we have an algorithm *S* which will compute $f(x_1, \ldots, x_n)$ if the arguments are expressed in notation *B*. We may give the following informal algorithm to compute *f* in notation *A*:

1. Given $x_1, \ldots, x_n$ in notation *A*, translate them into notation *B*, yielding $y_1, \ldots, y_n$.
2. Apply algorithm *S* to $y_1, \ldots, y_n$, giving $z = f(x_1, \ldots, x_n)$, in notation *B*.
3. Translate *z* back into notation *A*, giving $f(x_1, \ldots, x_n)$ in notation *A*.

PROPOSITION II.1.1

The class of all effectively computable functions on $N = \{0, 1, 2, \ldots\}$ is countable.

PROOF

Each such function must possess an algorithm for computing it. An algorithm is merely a string of symbols over an alphabet consisting of English letters and mathematical and punctuation symbols. We know that the set of all strings over a given alphabet is countable, so the set of all algorithms is countable; hence the set of all effectively computable functions must be countable as well.   ∎

1.2 Effectively Enumerable and Effectively Decidable Sets

In what way can we apply the idea of "effective process" to the problem of definition of sets? For example the set of prime numbers seems intuitively "effective," in that given an arbitrary number we can decide whether or not it is a prime; further, we can also devise a process which will list the set of primes, for example Eratosthenes' sieve. In practice a set can be defined in one of two ways: First, by *listing its elements*, either explicitly if finite (for example $S = \{2, 4, 7, 14\}$) or by giving a rule of enumeration (for example, S consists of the numbers 1^3, $1^3 + 2^3$, $1^3 + 2^3 + 3^3$, ..., for $n = 1, 2, 3, \ldots$); or second, by giving a *rule for membership* (for example, $S = \{n \mid n^2 + 1 \text{ is prime}\}$ defines the set which contains 1, 2, 4, 6, 10, 14, and so forth). Corresponding to this distinction we have two distinct measures of the effectiveness of a set:

1. A set S is **effectively enumerable** iff $S = \varnothing$, or there is an effective process which will list all and only the elements of S (possibly out of order or with repetitions). Thus if x is any object,

    (a) if $x \in S$, the process will eventually list x (perhaps after listing many others), and

    (b) if $x \notin S$, then x will never be listed.

2. A set S is **effectively decidable** iff there is an effective process which, when given an object x, will eventually answer "yes" if $x \in S$, and will eventually answer "no" if $x \notin S$. Thus the process will eventually halt for any input x.

For simplicity, we require the listing process to be one which produces a list of unbounded length. This is no restriction on the enumerability of finite sets, since such a set may be listed in an infinite way by repeating one of its elements. For example, 2, 4, 7, 14 may be enumerated as 2, 4, 7, 14, 14, 14, ...

Although these concepts are similar, and in fact every effectively decidable set is also effectively enumerable, we shall show that the converse is not true; that is, there are effectively enumerable sets which are not effectively decidable. This will be done informally in this chapter, and more formally (in terms of Turing machines) in Chapter V.

We assume without explicit statement that every set discussed is a subset of an effectively enumerable set U. In most cases U will be either the set of all strings over a finite alphabet A, or the set of nonnegative integers $N = \{0, 1, 2, \ldots\}$. Note that any effectively enumerable set is countable (this is immediate from the definition).

The following propositions establish the elementary properties of effectively decidable and enumerable sets. The effective processes which are used will be described by informal algorithms either as a series of imperative English statements, or by means of flow charts. If the process requires an input x, the flow chart will contain a box " READ x." If it produces an output y, the flow chart will contain a box " WRITE y."

However, not every flow chart or series of English statements defines an effective procedure. Two side conditions must be satisfied:

1. Each single statement or flow chart box must itself be effectively performable.
2. If the process is required to produce an output it must do so after some finite number of steps (that is, it may not go into an infinite "loop" if more output remains to be produced).

PROPOSITION II.1.2

A set $S \subseteq U$ is effectively enumerable iff $S = \varnothing$ or S is the range of an effectively computable total function $f : N \to U$.

PROOF

Clearly no problems arise if $S = \varnothing$, so suppose $S \neq \varnothing$. For the "if" part, let S be the range of f. Then the flow chart of Fig. II.1 defines an effective procedure to enumerate S.

For the converse, suppose the effective process P will list the elements of S. We may certainly assume that P will produce an infinite series of

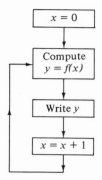

Figure II.1   Procedure to enumerate the range of *f*.

outputs, even if *S* is finite (by repeating the same element). Let $f: N \to U$ be defined so that for each $x \in N$, $f(x)$ equals the $(x + 1)$st element produced by *P*. Clearly *f* is total, and its range is *S*. Thus the proposition will be proved if we can show that *f* is effectively computable. This can be done by the flow chart of Fig. II.2.

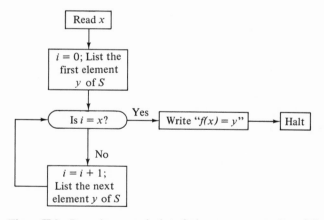

Figure II.2   Procedure to calculate *f*, given an enumeration of *S*.

It is easily seen that this process is indeed effective, terminates for all *x* in *N*, and computes *f*.   ∎

PROPOSITION II.1.3

If *S* is effectively decidable, then *S* is also effectively enumerable.

PROOF

Suppose effective process P will, when given any $x \in U$, answer "yes" or "no" according as $x \in S$ or $x \notin S$. Then the flow chart of Fig. II.3 defines an effective process which will enumerate all and only elements of S. Recall that U is assumed to be effectively enumerable. ∎

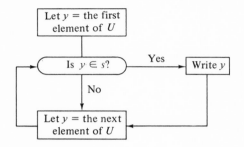

Figure II.3   Procedure to enumerate a decidable set S.

PROPOSITION II.1.4

A set $S \subseteq U$ is effectively decidable if and only if both S and $U\backslash S$ are effectively enumerable.

PROOF

Suppose S is effectively decidable. By the previous proposition S is effectively enumerable. Further, if the words "yes" and "no" are interchanged, the above flow chart will enumerate $U\backslash S$. This completes the "only if" part.

Now suppose that there is an effective process P which enumerates S, and an effective process Q which enumerates $U\backslash S$. We claim that the flow chart of Fig. II.4 will effectively decide whether its input x is in S. The cases $S = \varnothing$ and $U\backslash S = \varnothing$ are trivial, so we assume $S \neq \varnothing \neq U\backslash S$.

Clearly if this process terminates when given x, the answer produced is correct; further, if $x \in U$ then x must be in either S or $U\backslash S$, so that x will eventually be listed by either P or Q. Thus the process will terminate for any input in U. ∎

The proposition above establishes the exact correspondence between decidable and enumerable sets. However, we shall need considerably more machinery to show that there are enumerable sets which are not

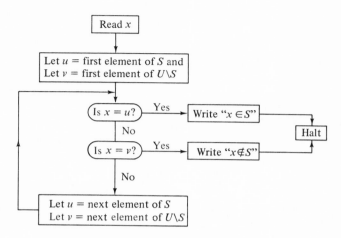

Figure II.4   Decision procedure for S.

decidable, since such a statement requires proof that there can be no effective procedure whatever to decide membership in these sets. In particular we need a more precise definition of effectiveness, which will be provided by the Turing machine as described in Section II.2.

The following shows that a set is effectively enumerable iff it is the domain of a computable partial function f; that is, the set of values x such that $f(x)$ is defined.

PROPOSITION II.1.5

If S is effectively enumerable, then S is the domain of an effectively computable partial function.

PROOF

We are given an effective procedure P which enumerates all and only the elements of S; we must produce an effectively computable function f, such that $f(x)$ is defined iff $x \in S$.

Let c be an arbitrary chosen element of U (e.g., 0 if $U = N$, or λ if $U = A^*$ for some alphabet A). Now define $f(x)$ as follows, for each $x \in U$:

$$f(x) = \begin{cases} c & \text{if} \quad x \in S, \\ \text{undefined} & \text{if} \quad x \notin S. \end{cases}$$

The flow chart of Fig. II.5 is an effective computing procedure for *f*.   ∎

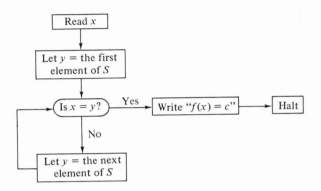

Figure II.5   A computable function *f* whose domain is an enumerable set *S*.

PROPOSITION II.1.6

A set *S* is effectively enumerable if and only if it is the domain of an effectively computable partial function.

PROOF

The previous proposition is the "only if" part. Hence, suppose that effective procedure *P* will compute the partial function *f*, and that $S \neq \varnothing$ is the domain of *f*. We must produce an effective procedure to enumerate *S*.

The natural first approach would be by a procedure such as the following:

1.   Let *x* = the first element of *U*.
2.   Calculate $f(x)$, using *P*.
3.   If $f(x)$ was defined, then write *x*.
4.   In any case, let *x* = the next element of *U* and go to step 2.

Unfortunately this procedure is not effective, since step 3 requires the completion of a possibly infinite process (since *P* will not terminate in case $f(x)$ is undefined). To get around this difficulty, recall that *P* must operate in a stepwise fashion. For simplicity we assume that $U = N$; the general case is similar.

We compute the values of $f(0), f(1), \ldots$ in the following way: First perform a single step of the computation of $f(0)$; then perform two steps in computing $f(0)$ and two steps in computing $f(1)$; then, three steps for $f(0)$, for $f(1)$, and for $f(2)$, and so forth. This process is made more precise in Fig. II.6.

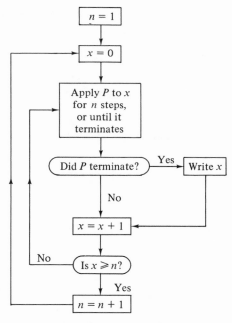

Figure II.6 Enumeration of the domain of a computable function.

This procedure is clearly effective, since each box is individually effective, and the fact that $S \neq \varnothing$ implies that it will produce an infinite sequence of output values. Let S' be the set that this procedure enumerates. We must show that $S = S'$. First, note that a number j is generated if the computation of $f(j)$ terminates. Thus j is in the domain of f, so $S' \subseteq S$. Conversely, suppose $j \in S$, and that P requires k steps to compute $f(j)$. In this case the test "did P terminate" will be answered "yes" when x has the value j, and $n = \max(j, k)$. At this point j will be generated, so $S \subseteq S'$. Thus $S = S'$ as required. ∎

Exercises

1. Show that any finite set is decidable.
2. Show that if a set is effectively enumerable, then it is effectively enumerable without repetitions.
3. Show that a set $S \neq \varnothing$ can be effectively enumerated in increasing order if and only if S is decidable.
4. Show that if S is the range of an effectively computable partial function, then S is effectively enumerable.
5. In what order does the algorithm of Proposition II.1.6 enumerate the domain of a partial function g?
6. Show that if S and T are effectively enumerable (effectively decidable), then $S \cup T$ and $S \cap T$ are of the same type.
7. Show that an infinite effectively enumerable set must contain an infinite effectively decidable subset. *Hint:* Use Exercise 3.

1.3 Uncomputable Functions, Undecidable Sets, and Halting Problem

In this section we will (informally) prove the existence of functions which are finitely describable but not effectively computable. The technique used is that of diagonalization applied to an effective enumerating function for the set of all unary effectively computable functions. A great number of the negative results of the theory of computability are based on similar arguments (but of course in more formal contexts). The reader may now wish to review the proof of Theorem I.5.5, and the definition of an enumerating function for a class of functions.

The following shows that uncomputable functions must exist.

PROPOSITION II.1.7

There are total functions on N which are not effectively computable.

PROOF

By Theorem I.5.5, the set S of all functions $f: N \to N$ is not a countable set. However Proposition II.1.1 states the set of functions f in S which are effectively computable is countable. Hence there is an uncountable number of uncomputable functions (for otherwise S would be countable by Exercise 4 of Section I.5.1).   ∎

More concretely, the following proposition shows how we can obtain an uncomputable function from any enumeration of all the computable total functions.

PROPOSITION II.1.8

Let $U(k, x)$ be any function which enumerates the set T, where

$$T = \{f : N \to N \mid f \text{ is a computable total function}\}.$$

Then U cannot be an effectively computable function.

PROOF

Clearly U must be total. By definition of the enumeration function U, a function $f : N \to N$ is in T if and only if for some fixed k, we have

$$f(x) = U(k, x) \qquad \forall x \in N.$$

Now suppose that U is effectively computable. Let the function $g : N \to N$ be defined as follows, for each $x \in N$:

$$g(x) = U(x, x) + 1.$$

Clearly g is computable, by virtue of the assumption that U is computable. Further, g is total, so g satisfies the conditions for membership in T. Consequently, there must be a k_0 such that $g(x) = U(k_0, x)$ for all x. But this implies: $U(k_0, k_0) = g(k_0)$ by substitution and $g(k_0) = U(k_0, k_0) + 1$ (by definition of g).

Thus $0 = 1$, which is impossible. Hence the assumption that U is effectively computable must be incorrect. ∎

The next proposition states that the set of computable *partial* unary functions *is* effectively enumerable. The argument we give is necessarily imprecise, since we do not yet have a precise, exact definition of effective process or algorithm. However, the reasoning is exactly the same as will be used later in a formal context.

PROPOSITION II.1.9

There is an effectively computable partial function U which enumerates the set of all effectively computable partial unary functions.

Every effectively computable function must possess an algorithm which can be used to compute it. An algorithm is a string of symbols over an alphabet A which consists of the English alphabet, together with the numerals $0, 1, \ldots, 9$ and various mathematical and punctuation symbols. Thus the set L of all algorithms is a subset of the set A^* of all words over A.

Further, we can place restrictions on the format in which an algorithm is expressed, so that it becomes a simple matter to determine whether an arbitrary word x in A^* is in fact an algorithm. For example, we might without loss of generality require that an algorithm must be expressed as a sequence of statements numbered $1, 2, 3, \ldots$ by consecutive integers. In this case, each instruction would specify an effective operation to be performed. In addition we can specify that each such statement is of a very simple, atomic nature. This could be achieved by breaking up each single complex instruction into a sequence of simpler instructions.

As a consequence the set $L \subseteq A^*$ of algorithms expressed in our restricted format will be an effectively decidable set of words (the validity of this statement will be much more evident once we have a more precise formulation of " algorithm "). A^* is itself effectively enumerable in some fixed order, so we can effectively list the set L as $l_0, l_1, l_2, \ldots$ so that if $i < j$, then l_i is earlier in the enumeration of A^* than is l_j.

We define the function U by:

$$
U(k, x) = \begin{cases} y & \text{if the } k\text{th algorithm } l_k \text{ will eventually halt and produce } y, \text{ when applied to } x, \\ \text{undefined} & \text{if } l_k \text{ will not eventually halt when applied to } x. \end{cases}
$$

This function can be computed effectively, as by the flow chart of Fig. II.7.

In this flow chart, each box is clearly performable in an effective manner, with the possible exception of the one containing "apply l_k to x." Recall that by definition an algorithm is a finite, *executable* description of an effective process; thus it is possible to perform the sequence of steps which l_k specifies, in an effective manner.

Only two outcomes are possible: either this process will terminate, in which case $U(k, x)$ is defined and equal to the result produced by l_k

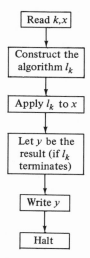

Figure II.7 Procedure to calculate $U(k, x)$.

applied to x; or it will never terminate, in which case $U(k, x)$ is undefined. Thus $U(k, x)$ is effectively computable.

Now we must show that U in fact enumerates the set of all effectively
computable partial unary functions. Call this set T, and denote by T'
the set of functions enumerated by U.

First, U is an effectively computable function of two arguments.
Thus, for any fixed $k_0 \in N$, the function

$$f(x) = U(k_0, x)$$

of a single variable x is also effectively computable. Thus $T' \subseteq T$.

Now let f be any function in T; then f is effectively computable, so
it must possess an algorithm l_{k_0} for computing its values. Thus for
each x, $f(x)$ is (if defined) the result of applying l_{k_0} to x, and $f(x)$ is
undefined in case l_{k_0} will not terminate when applied to x. But this
implies by our construction of U, that

$$f(x) = U(k_0, x) \qquad \forall x \in N.$$

Thus $f \in T'$, so $T \subseteq T'$ and consequently $T = T'$. ∎

REMARKS

The argument of Proposition II.1.8 does not apply in this case, for
the following reason. That argument proceeded by defining $g(x)$ to

equal $U(x, x) + 1$ for all $x \in N$, and observing that the assumption that U is computable must imply that g is also computable, and so in the enumeration. Consequently for some fixed k_0, $g(x) = U(k_0, x)$ for all x.

This portion of the argument also applies in the current context. However, the climax of the argument of II.1.8 arose from the equation

$$U(k_0, k_0) = g(k_0) = U(k_0, k_0) + 1 \qquad (1)$$

which cannot be satisfied by any total function g, since it implies that $0 = 1$. However in the current situation U *is a partial function*, so that Equation (1) can be satisfied by virtue of the fact that $U(k_0, k_0)$ is undefined.

A second remark is that Proposition II.1.8 and II.1.9 together imply the perhaps surprising fact that a set may be effectively enumerable, while a subset of it is not effectively enumerable. While this may seem intuitively implausible, analogous situations often occur among decidable sets. For example, the set P of prime numbers greater than 2 is a subset of the set Q of all the odd integers. However it is clearly much easier to enumerate Q than it is to enumerate its subset P.

Finally, the least concrete portion of the argument of Proposition II.1.9 is the idea of expressing all possible algorithms as a decidable set $L \subseteq A^*$. In the more formal part of this book, the term "effectively computable" will become "computable by a Turing machine," and the term "algorithm" will become "Turing machine program." In that context, the decidability of L becomes much more transparent.

An Uncomputable Function and the Halting Problem

The **halting problem** is the following: given an algorithm z and an input x, to decide whether or not z will eventually terminate when applied to x.

PROPOSITION II.1.10

The halting problem is not effectively decidable.

PROOF

Suppose, for the sake of argument, that an effective procedure does exist to decide the halting problem. Then this procedure could certainly be applied to any of the algorithms l_k ($k = 0, 1, 2, \ldots$) as used in Proposition II.1.9. Now define a function g so that for any $x \in N$,

$$g(x) = \begin{cases} 0 & \text{if } l_x \text{ will not terminate when applied to } x \\ \text{undefined} & \text{if } l_x \text{ will terminate when applied to } x. \end{cases}$$

Consider the flow chart of Fig. II.8.

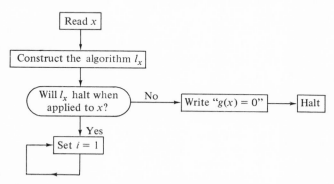

Figure II.8 Procedure to compute g.

Clearly this flow chart describes a process for computation of g. Furthermore, our assumption that the halting problem is effectively decidable implies that this process is effective, so g is effectively computable. We shall show that this leads to a contradiction.

Since g is computable, the algorithm described by the flow chart must appear in the list $l_0, l_1, l_2, \ldots$, say at position k_0. Consider the value of $g(k_0)$; that is, the result of applying the algorithm for g to its own index in the list $l_0, l_1, l_2, \ldots$. Only two cases are possible: either $g(k_0)$ is defined or it is undefined.

If $g(k_0)$ is defined, then the flow chart must take the branch leading to the "halt" box (for otherwise its computation would not terminate). However, this implies that the answer to the test "will l_{k_0} halt if applied to k_0?" must be "no"; consequently by the definition of g, it must be the case that $g(k_0)$ is undefined, a contradiction.

As a result it can only be true that $g(k_0)$ is undefined. However this can occur only if the branch leading to the box "set $i = 0$" is taken, and this branch can be taken only if $g(k_0)$ is defined, another contradiction.

Thus the assumption that the halting problem is effectively decidable implies that g is effectively computable, which in turn leads to an impossible situation. Thus the halting problem is not effectively decidable.   ∎

COROLLARY II.1.11

There are finitely describable functions which are not effectively computable.

PROOF

Function g cannot be effectively computable, as was just shown. However g is completely (and finitely) defined in that proof.   ∎

This contradicts the often-stated claim that "if a process can be described precisely enough, it can be programmed."

As another consequence we obtain the following:

PROPOSITION II.1.12

There are effectively enumerable sets which are not effectively decidable.

PROOF

Define $S \subseteq N$ by

$$S = \{x \mid l_x \text{ terminates when applied to } x\}.$$

Now S is clearly the domain of the 1ary partial function $f(x) = U(x, x)$. Consequently (by Proposition II.1.6), S is effectively enumerable.

If S were effectively decidable, we could compute the function g of Proposition II.1.10 as in Fig. II.9.

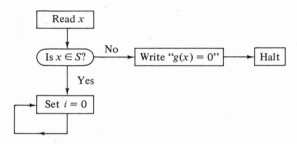

Figure II.9   Procedure to compute g.

Since g is not effectively computable, S cannot be effectively decidable.   ■

Reduction of Other Problems to the Halting Problem

Suppose P is some general problem which we wish to prove is not effectively solvable. This can be done by establishing the following implication, which leads to a contradiction of Proposition II.1.10:

> If there were an effective process for solving P, then we could use this to obtain an effective process which would solve the halting problem.

As an application of this technique, let us define the *printing problem* to be the following: given an algorithm z which contains an instruction "print the symbol $*$," and an input x, the problem is to determine whether the instruction to print "$*$" will ever be executed during the computation by z on x.

It is not necessary to solve the halting problem in order to solve the printing problem, since it is easy to construct algorithms which print $*$ and do not halt, or do not print $*$ but do halt, or do both or neither. The following, however, is a valid argument which will be formally paraphrased in Chapter V.

Suppose the printing problem were decidable. Then it could be used to solve the halting problem as follows:

1. Given an algorithm z and an input x, construct a new algorithm z' as follows:
2. Let "$*$" be any symbol which does not appear in z, so z cannot print it.
3. Modify z to obtain z', so that z' will, when given x
   (a) Perform the operations of z;
   (b) Print "$*$" just in case z halts, given x.
   Note: This can be done effectively; for example, replace any "halt" instruction in z by the two instructions: "print $*$; halt."
4. Decide whether z' will print "$*$" when given x; if so answer "z halts if given x"; if not, answer "z will not halt if given x."

Clearly, if given x, z' will print "$*$" if and only if z will halt. Thus this "algorithm" effectively decides the halting problem, which contradicts Proposition II.1.10. Thus the printing problem is also not effectively decidable.

Exercises

1. Show that the following is not effectively decidable: Given an algorithm z which computes a unary function f, and given an input x, the problem is to decide whether or not $f(x) = \lambda$.
2. Let $f, g: (A^*)^n \to A^*$ be effectively computable such that g is total. By definition g is a *time bound* for f iff the following holds:

    (a) there is an effective procedure P for f such that
    (b) for all $\vec{x}_n \in A^*$, if $f(\vec{x}_n)$ is defined, then procedure P will, when given $\vec{x}_n$ as input, terminate within $g(\vec{x}_n)$ steps; and
    (c) if $f(\vec{x}_n)$ is undefined then P will never terminate.

    Prove that the enumerating function $U(k, x)$ of Proposition II.1.9 does not possess an effectively computable time bound.
3. Prove that there is a unary function $f: A^* \to A^*$ which has no effectively computable time bound.

2. TURING MACHINES—PRELIMINARY DEFINITIONS

In order to give the results of the previous section a more rigorous foundation, it is necessary to adopt a precise definition of the terms "effective process," "effectively decidable," and so forth. This will be done by means of a simple, abstract computing device called the "Turing machine." In particular the results of the previous section (called "propositions") will all be seen to have close analogs in terms of "Turing decidable," "Turing computable," and so forth.

In the large, the purpose of this book is dual:

1. To present various arguments to the effect that the Turing machine is an adequate, complete model of intuitively effective processes.
2. To establish the basic properties of the effectively computable functions, and the effectively enumerable and decidable sets.

The first point just mentioned is known as the Church–Turing thesis, and will be discussed in more detail at the end of this chapter. If we grant the validity of this thesis, it follows that many results which apply to Turing machines have a larger scope as well. For example, a precisely defined function which is not Turing computable must then be not computable by any effective process whatever.

In this section we give a description of the structure of the Turing machine and the way it is used to compute functions and accept sets or predicates. Turing's original papers (for example [T1]) were written in the 1930s in connection with the Entscheidungsproblem (the problem of the effective decidability of predicates). The formalism of Turing machine and the approach to computability in these notes is closely related to Turing's original formulation in spirit, although differing somewhat in details. Our version of the Turing machine is essentially that of Wang [W1]. Equivalence of the various definitions will be discussed at the end of this section. In any case, one of our main results will be that the concept of computability is invariant under many changes in the basic computing devices, so the exact formalism chosen is not important.

2.1 Turing Machines

A concise, strictly formal definition of Turing machines and related topics will be given in the next chapter. The current definitions are sufficiently precise for ordinary mathematical purposes, but a completely formalized definition will be required later, since the Turing machines themselves will be objects of mathematical study.

Informally, a Turing machine consists of a *tape*, a *read–write head*, and a *program*. The tape is infinite and one dimensional, and is divided into a sequence of squares, each capable of holding any symbol from an alphabet A, or a special symbol $*$ (not in A) which plays the role of a blank. Any tape which we consider will contain only a finite number of nonblank squares. The read–write head is a device which at any moment is scanning a single square of the tape, and will obey one of the program's instructions in a single step. An instruction can cause it to write a new symbol on the square it is scanning, thus erasing the symbol previously there; the head may move itself one square to the right, or to the left, thus scanning an adjacent square; or it may read the symbol on the square it is scanning (without changing the symbol) and, on the basis of the symbol it found, choose one of two alternate instructions to obey at the next step.

An important point: the Turing machine tape, while conceptually infinite in length, will only contain a finite number of nonblank squares at any one time. Thus it can be described finitely, by specifying only the nonblank portion; all remaining squares are assumed to contain blanks. However, the nonblank portion may grow or shrink during a computation (Fig. II.10).

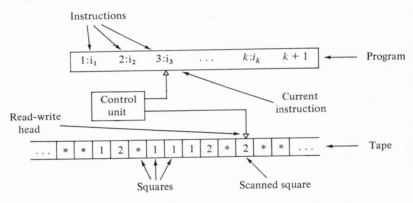

Figure II.10   Intuitive concept of a Turing machine.

The actions of the Turing machine are determined by its program.

DEFINITION II.2.1

A **program** is a finite sequence of instructions, each with a unique label p_i, which ends with a label. Thus a program P is a word of the form $p_1 : i_1\ p_2 : i_2 \cdots p_k : i_k\ p_{k+1}$. Each instruction i_j has one of the forms:

1. R (denoting " move the read–write head right one square "),
2. L (denoting " move the read–write head left one square "),
3. a (denoting " write a on the scanned square "),
4. Jq (denoting " jump to the instruction with label q "), or

5. *Jaq* (denoting "jump to the instruction with label q if scanning the symbol a").

In these, a must be either a symbol of A or the blank symbol $*$, and q must be the label of some instruction in the program. ∎

The Turing machine normally obeys the instructions in the order that they occur in P, so that after performing i_j it then proceeds to i_{j+1}, and halts after performing i_k. However, this sequence may be broken by an instruction of the form "Jap_l." If the machine is scanning a, the next instruction is that with label p_l; if not scanning a, execution proceeds to i_{j+1} as usual. The instruction Jp_l causes the next instruction to be that with label p_l, regardless of the symbol scanned.

For an example, consider the Turing machine Z_1 with alphabet $A = \{1, 2\}$ and program,

$$p_1 : J*p_8 \quad p_2 : J1p_5 \quad p_3 : 1 \quad p_4 : Jp_6$$
$$p_5 : 2 \quad p_6 : R \quad p_7 : Jp_1 \quad p_8$$

If Z_1 is given the tape $\cdots * 1\ 2\ 1\ * 2\ * \cdots$ and is initially scanning the leftmost 2, then Z_1 will perform the series of steps shown in Table II.1

Z_1 will interchange the digits 1 and 2, beginning with the square initially scanned and continuing to the right until the symbol $*$ is found. For another example, we give a machine Z_2 with the same alphabet which will convert a dyadic integer x into $x + 1$. Z_2 begins at the right end of x. P_2 (the program of Z_2) is given by

$$P_2 = p_1 : J1p_8 \quad p_2 : J*p_6 \quad p_3 : 1 \quad p_4 : L \quad p_5 : Jp_1$$
$$p_6 : 1 \quad p_7 : Jp_9 \quad p_8 : 2 \quad p_9$$

Z_2 will convert 121121 into 121122, 121122 into 121211, and 222 into 1111.

Computation of Functions

Suppose Z is a Turing machine over alphabet A, and that $P = p_1 : i_1 \cdots p_k : i_k \quad p_{k+1}$ is its program. If $(x_1, \ldots, x_n)$ is any n-tuple of words from A^*, the tape $\cdots ** x_1 * x_2 * \cdots * x_n ** \cdots$ is called the **initial tape** corresponding to $(x_1, \ldots, x_n)$. Z is in the **initial configuration** for

TABLE II.1

Current Instruction p_i	Effect of p_i	Nonblank Portion of Tape and Scanning Position Just before Executing Instruction p_i
p_1	Test for *	1 2 1 * 2 ↑
p_2	Test for 1	1 2 1 * 2 ↑
p_3	Write 1	1 2 1 * 2 ↑
p_4	Jump to p_6	1 1 1 * 2 ↑
p_6	Move right	1 1 1 * 2 ↑
p_7	Jump to p_1	1 1 1 * 2 ↑
p_1	Test for *	1 1 1 * 2 ↑
p_2	Test for 1	1 1 1 * 2 ↑
p_5	Write 2	1 1 1 * 2 ↑
p_6	Move right	1 1 2 * 2 ↑
p_7	Jump to p_1	1 1 2 * 2 ↑
p_1	Test for *	1 1 2 * 2 ↑
p_8	Halt	1 1 2 * 2 ↑

$(x_1, \ldots, x_n)$ if it is scanning the leftmost symbol of $x_1 * \cdots * x_n *$, and is about to execute instruction i_1.

This describes the standard input format for a Turing machine. We do not specify an output format rigidly but instead give a rule by which any configuration of Z may be interpreted as containing an "output" word from A^*. Suppose the tape of Z is $\cdots * * a_1 a_2 \cdots a_m * * \cdots$, where each $a_i \in A \cup \{*\}$; suppose further that Z is currently scanning symbol a_k. Then the **content** of this configuration is the longest word $a_k a_{k+1} \cdots a_l$ such that $a_k, a_{k+1}, \ldots, a_l$ all lie in the alphabet A; all remaining symbols $(a_1, \ldots, a_{k-1}$ and $a_{l+1}, \ldots, a_m)$ are ignored for this purpose.

Thus the content of a configuration α is the longest word y in A^*, such that y is part of α and the leftmost symbol of y is being scanned. Note that y must equal λ if the symbol scanned in α is $*$. Examples appear in Table II.2.

TABLE II.2

Tape and Square Scanned	Content
$\cdots * 1\ 2\ 1 * \cdots$ $\uparrow$	121
$\cdots * * * \cdots$ $\uparrow$	λ
$\cdots * 1\ 2\ 1\ 2 * 1 * \cdots$ $\uparrow$	212
$\cdots * 2 * 1\ 2 * 1\ 1\ 2 * \cdots$ $\uparrow$	2
$\cdots * 2 * 1 * * 1\ 1\ 2 * \cdots$ $\uparrow$	λ

DEFINITION II.2.2

Let $f : (A^*)^n \to A^*$ be a partial $n$ary function over A^*. By definition f is **Turing computable**, or **recursive** if there is a Turing machine Z such that for all $x_1, \ldots, x_n, y \in A^*$

(i) if $f(x_1, \ldots, x_n) = y$ and Z is started in the initial configuration for $(x_1, \ldots, x_n)$, then Z will obey its instructions and eventually halt with a tape whose content is y;

(ii) if $f(x_1, \ldots, x_n)$ is undefined, then Z will never halt after being started in the initial configuration for $(x_1, \ldots, x_n)$. ∎

Machines Z_1 and Z_2 (given earlier as examples) do not compute interesting functions, but only because neither halts at the left of its result. Thus Z_2 will compute the function $f(x) = x + 1$, if it is modified to first find the right end of its input, then to act as given before and finally to find the left end of the resulting word over $A = \{1, 2\}$.

REMARKS

By this definition it is easy to see that if Z is any Turing machine over alphabet A, and if $n > 0$, then Z computes an $n$ary partial recursive function Ψ_Z. In fact, $\Psi_Z(\vec{x}_n)$ can be as follows. Suppose that Z is given $(x_1, \ldots, x_n) \in (A^*)^n$ in the initial configuration format. Then

$$\Psi_Z(\vec{x}_n) = \begin{cases} y & \text{if } Z \text{ will eventually halt in a con-} \\ & \text{figuration whose content is } y, \\ \text{undefined} & \text{if } Z \text{ does not halt.} \end{cases}$$

In general, the class of all partial recursive functions can be enumerated by simply enumerating all possible Turing machine programs.

Recursive and Recursively Enumerable Sets

DEFINITION II.2.3

A set $S \subseteq A^*$ is **recursively enumerable** iff there is a Turing machine Z, such that if Z is started with a word $x \in A^*$ on its tape in its initial configuration, then

   (i) Z will eventually halt if $x \in S$ and
  (ii) Z will not halt if $x \notin S$.

A set $S \subseteq A^*$ is **recursive**, or **Turing decidable** iff there is a Turing machine Z such that, for any $x \in A^*$, if Z is started in its initial configuration scanning x, then

   (i) Z will eventually halt, scanning a blank symbol $*$ if $x \in S$ and
  (ii) Z will eventually halt, scanning a nonblank symbol if $x \notin S$. ∎

Note that the word "recursive" is used in two distinct contexts, one involving functions and the other involving sets. Clearly S is recursively enumerable iff it is the domain of a recursive partial function.

Recursive and Recursively Enumerable Predicates

DEFINITION II.2.4

A predicate $P(x_1, \ldots, x_n)$, whose variables represent words over A, is **Turing decidable**, or **recursive** iff there is a Turing machine Z such that for all $x_1, \ldots, x_n \in A^*$, if Z is started in its initial configuration, then either

(i) $P(x_1, \ldots, x_n)$ is true and Z eventually halts, scanning the symbol $*$ or

(ii) $P(x_1, \ldots, x_n)$ is false and Z eventually halts, scanning a symbol from A.

P is said to be **recursively enumerable** iff there is a Turing machine Z as above, such that for each $x_1, \ldots, x_n \in A^*$, either $P(x_1, \ldots, x_n)$ is true and Z halts, or $P(x_1, \ldots, x_n)$ is false and Z never halts. If P is recursive as above, we say that Z **decides** the predicate P; in the other case we say that Z **accepts** P. Clearly the case $n = 1$ reduces to the definition already given for sets. ∎

Notice that, by this definition, any Turing machine Z accepts recursively enumerable predicates $P_n(\vec{x}_n)$, for $n = 1, 2, \ldots$, so that $P_n(\vec{x}_n)$ is true for all and only those $(\vec{x}_n) \in (A^*)^n$ such that $\Psi_Z(\vec{x}_n)$ is defined.

2.2 Other Formulations of the Turing Machine

Several variants of the Turing machine have appeared in articles on computability. Turing's original version [T1], for example, has a one-way infinite tape, and is directed by a set of finite set of quintuples rather than a program. This type of machine possesses a finite number of *states* $q_0, q_1, \ldots, q_k$. At any instant the next action is determined by the state the machine is in and the symbol it is scanning. Its actions each consist of three steps: write a symbol, then move the read–write head, and then change the state. All this is specified by the quintuple (q_i, a, b, m, q_j), where a and b are tape symbols (in $A \cup \{*\}$), a is the symbol being scanned, b is the symbol to be written, m is R, L, or C (indicating a move of one square to the right, to the left, or not at all), q_i is the current state, and q_j is the next state.

A *Turing machine* is, by this definition, a finite set of quintuples such that no two begin with the same state and tape symbol. (This ensures that the machine is deterministic.) If the machine ever reaches a state-symbol pair which begins none of its quintuples, it halts.

The Post machine [P3] is very similar to Turing's formulation, but uses quadruples instead of quintuples. A quadruple is of the form (q_i, a, x, q_j), where x is either R or L, or a tape symbol. Thus a Post machine may either move its read–write head, or it may write, but not both in the same step. It also must be deterministic.

Our version is very similar to the Wang machine [W1]. The choice of control by program, rather than by quadruples or quintuples was made for one reason: the program format increases the Turing machine's similarity to a digital computer. This means that some programming concepts such as *subroutines* and *flow charts* can be borrowed directly to be used in constructions, and brings the concept of Turing computability slightly closer to practical computability.

It is easy to see that the programs and sets of quintuples are equivalent. All that is necessary is to show two things:

(i)   For any program P it is possible to construct a set of quintuples S which have the same effect;

and

(ii)   for any set of quintuples S, it is possible to construct an equivalent program P.

To prove (i) is very easy: If P contains $p : R\,q$, $p : L\,q$, or $p : a\,q$, then S contains (p, b, b, R, q), (p, b, b, L, q) or (p, b, a, C, q) for each $b \in A \cup \{*\}$. If P contains $p : Jaq\,r$, then S contains (p, a, a, C, q), and (p, b, b, C, r) for each $b \in A \cup \{*\}$ such that $b \neq a$. Clearly these quintuples merely duplicate the effect of the instructions of P.

The proof of (ii) is equally straightforward, but the details are messier. The basic idea is to associate with each state q_i a series of instructions which will determine which symbol is being scanned, and will then apply the appropriate quintuple.

Thus differences in instruction arrangement and format give little trouble. We may also have differences in memory or storage structure, such as a one-way infinite tape versus a two-way infinite tape, or one tape versus five tapes, or a one-dimensional tape versus a two- or higher-dimensional tape. A natural question, in view of the extreme simplicity of the Turing machine is this: Wouldn't it be possible to give Turing machines more computational power by adding extra capabilities?

The answer to this question seems to be invariably no. While these features may make computations faster, they do not increase the class of all computable functions, predicates, and so on, as long as the extra powers can be simulated by a Turing machine. For example, a multi-dimensional tape can be uniquely represented by an ordinary tape (much longer) in an effective way. Thus an ordinary Turing machine can simulate such a "supermachine" by first converting its input to a one-dimensional form, and then simulating each of the actions of the supermachine on its n-dimensional tape by means of a corresponding series of actions on the one-dimensional tape.

Exercises

1. Construct Turing machines to compute the following functions. Assume the alphabet is $A = \{1, 2\}$.

    (a) $f(x) = x121$;
    (b) $f(x, y) = xy$;
    (c) $f(x, y) = x + y$.

2. Define $S = \{x \mid$ the length of x is even$\}$. Prove that S is a recursive set by constructing a Turing machine to decide S. Again, assume that $A = \{1, 2\}$.
3. Give a value of y for which $\Psi_Z(x_1, \ldots, x_n, y) = \Psi_Z(x_1, \ldots, x_n)$ is always true.
4. Prove that every recursive set is also recursively enumerable.
5. Describe a reasonable set of instructions for a Turing machine with a two-dimensional tape. How would you define computability of functions from $(A^*)^n \to A^*$?
6. Design a scheme which will make it possible to represent every two-dimensional tape uniquely as a one-dimensional tape. Is your scheme effective?
7. Describe, in English, a method by which the two-dimensional Turing machine can be simulated by an ordinary Turing machine.
8. Show that it is possible for a Turing machine to get into a computation which will never halt, but which is not repetitive; that is, it will never repeat the same combination of tape content and current instruction. Give an explicit Turing machine.

3. UNIVERSAL TURING MACHINES AND THE HALTING
PROBLEM

Note: The material in this section is not necessary for later chapters; in fact these results will appear again in Chapter V as a consequence of the rather complex development of Chapters III and IV. However the techniques used here are considerably simpler than those of the later chapters, and will certainly be of interest to students who do not plan to ge beyond the end of this chapter.

The first subsection contains a direct proof that the halting problem for Turing machines cannot be Turing decidable, and exhibits a finitely describable function which cannot be Turing computable. This construction is at a very elementary level, not requiring the concept of an enumerating function.

The second subsection outlines the construction of a universal Turing machine, which proves that the class of partial recursive unary functions over $A = \{1, 2\}$ is effectively enumerable. This can be used (as in Section II.1.3) to prove the undecidability of the halting problem for Turing machines.

3.1   The Halting Problem for Turing Machines

The following construction is essentially due to Rado [R1], as modified by Dunham [D2]. It gives a very direct proof of the undecidability of the halting problem for Turing machines, without bringing in the concept of effective enumerability of a set of functions. The crucial point is the observation that for each n, there is only a finite number of Turing machines over a one-symbol alphabet $A = \{1\}$, which have exactly n instruction labels $p_1, p_2, \ldots, p_n$.

Notation: For each $n > 0$,

1. T_n is the collection for all Turing machines with input alphabet $A = \{1\}$ which have exactly n instruction labels $p_1, \ldots, p_n$ (in that order);

2. if Z is a Turing machine which is started on a tape containing 1^n, then

$$f_Z(n) = \begin{cases} m & \text{if } Z \text{ will eventually halt, with tape} \\ & \text{content } 1^m; \\ \text{undefined} & \text{if } Z \text{ will not halt}; \end{cases}$$

3. H_n is the collection of all Turing machines in T_n such that $f_Z(n)$ is defined;
4. (a) $\varphi(0) = 0$,
    (b) if $n > 0$, then $\varphi(n) = m + 1$, where
    $m = \text{maximum } \{f_Z(n) \,|\, Z \in H_n\}$.

ASSERTION

The set T_n is effectively obtainable from n, and is finite for each $n > 0$.

ASSERTION

For each n, H_n is finite.

REASON

$H_n \subseteq T_n$ and T_n is finite. (However there is no obvious effective way to select the members of H_n from T_n.)

ASSERTION

φ is a well-defined total function.

REASON

Clearly any given Turing machine is either in H_n or it is not (even if we may not be able to decide which), since it either eventually halts or computes forever when given 1^n as input. Thus H_n is well defined. Consequently the maximum used to define φ ranges over a well-defined finite set, so for each n, $\varphi(n)$ is well defined. Totality of φ is immediate.

THEOREM II.3.1

φ is not Turing computable.

PROOF

Suppose that Turing machine Z_φ computes φ. We may assume that the instruction labels of Z_φ are $p_1, p_2, \ldots, p_k$ (in this order), by renaming the labels if necessary; $k - 1$ is the number of instructions in the program of Z_φ.

Since by assumption Z_φ computes φ, it follows that if Z_φ is given an input of the form 1^k, Z_φ will eventually stop with its tape containing $\varphi(k) = m + 1$ ones, where

$$m = \text{maximum}\{f_Z(k) \mid Z \in H_k\}.$$

Now Z_φ must itself be in H_k, since it eventually halts for all inputs (since φ is total). Thus $f_{Z_\varphi}(k) = \varphi(k)$ is among the set whose maximum is being calculated, so $m \geq \varphi(k)$. But this is impossible, since by definition $\varphi(k) = m + 1$ (since $m \geq \varphi(k) = m + 1$ implies $0 \geq 1$). Thus the assumption that φ is Turing computable must be false. ∎

COROLLARY II.3.2

The halting problem is not Turing decidable.

PROOF

Suppose it were Turing decidable. Then the informal algorithm of Figure II.11 could be used to compute φ. Each box in this flow chart is clearly effective, either because it is very simple or by the assumption that the halting problem is decidable. To complete the proof in full detail it is necessary to show that the algorithm above could be done by a Turing machine. This is left to the reader to verify (or he may wish to circumvent the problem by use of the Church–Turing thesis). ∎

3.2 A Universal Turing Machine

DEFINITION II.3.3

Let $U(k, x)$ be any function which enumerates the set of all partial recursive unary functions over $A = \{1, 2\}$. A **universal Turing machine** is any Turing machine which computes U.

Thus a universal Turing machine is presented with two arguments: a string $k \in A^*$, which specifies which function in U's enumeration

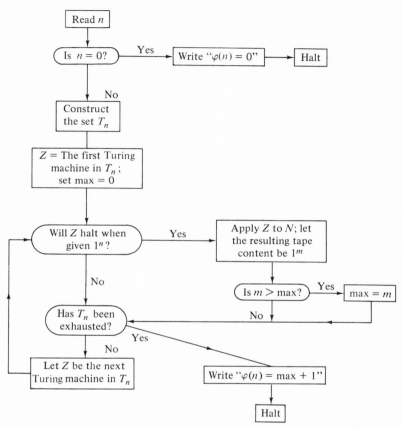

Figure II.11   Procedure to calculate $\varphi(n)$.

$f_0, f_1, f_2, \ldots$ is to be applied; and an argument $x \in A^*$ to which f_R is to be applied.

The most natural way to obtain such an enumeration function U is to let k be a word which represents the program P of a Turing machine Z; computation of $U(k, x)$ then proceeds by simulating the behavior of Z when applied to x. We shall sketch the construction of a Turing machine which computes U.

First, it is necessary to find a way to represent the program P of an arbitrary Turing machine Z over $A = \{1, 2\}$ by means of a coded word $\bar{P}$ over A, in such a way that $\bar{P}$ can be decoded to obtain P again.

This can be done by simply coding each symbol which can occur in a program $P = p_1 : i_1 \, p_2 : i_2 \cdots p_k : i_k \, p_{k+1}$ as a string of the form

Symbol:	R	L	J	$:$	$*$	1	2	$p_i \, (i = 1, 2, \ldots)$
Code:	212	$21^2 2$	$21^3 2$	$21^4 2$	$21^5 2$	$21^6 2$	$21^7 2$	$21^{i+7} 2$

It is easily verified that this code is effectively and uniquely decodable. If α is any string over the alphabet $\{R, L, J, :, *, 1, 2, p_1, p_2 \ldots\}$, we shall let $\bar{\alpha}$ denote the result of encoding each symbol of α as above.

While our universal Turing machine is in the process of simulating Z, it will have a tape of the general form: $\bar{P} * \bar{u} \; \bar{p}_i \; \bar{v}$. This represents the program P of Z, the instruction p_i which Z is about to execute, and the current tape contents uv of Z. The symbol which Z is scanning is the leftmost symbol of v.

We are now ready to describe the mode of operation of this universal Turing machine. It is left to the reader to fill in the details, that is, showing that each step below can indeed be performed by a Turing machine.

1. The initial configuration will be of the form $k * x$, where k, $x \in A^*$.
2. Check to see whether k is equal to $\bar{P}$ for some program P. If not, go into a never-ending loop.
3. If so, then encode x, and change the tape contents to: $\bar{P} * \bar{p}_1 \; \bar{x}$. This represents the initial configuration of Z.
4. Let the current tape contents be $\bar{P} * \bar{u} \; \bar{p}_l \; \bar{v}$ (initially, $u = \lambda$, $l = 1$, and $v = x$);

    (a) locate the lth instruction $\bar{i}_l$ in $\bar{P}$;
    (b) simulate the effect of i_l upon tape uv, obtaining a new configuration $\bar{P} * \bar{u}_1 \; \bar{p}_j \; \bar{v}_1$.

5. If p_j is not the last instruction label in P, then go to Step 4.
6. Otherwise do the following:

    (a) erase $\bar{P} * \bar{u}_1 \; \bar{p}_j$, yielding $\bar{v}_1$;
    (b) decode $\bar{v}_1$, yielding v_1;
    (c) halt.

It should be clear that this process will simulate Z one step at a time, and will yield the same output that Z yields when given x. Consequently we have two theorems:

THEOREM II.3.4

The halting problem for Turing machines is not Turing decidable.

THEOREM II.3.5

There is a set which is recursively enumerable but not recursive (that is, not Turing decidable).

Proofs are obtained from the proofs of Propositions II.1.10 and II.1.12 by replacing their constructions involving flow charts by constructions involving Turing machines.

4. PURPOSE, SIGNIFICANCE, AND PLAN

4.1 Purpose

The chief purpose of this book is to establish the basic properties, powers, and limitations of effective processes. A case in point is the class of all effectively computable functions—what are the limits (if any) of this class, what can cause a function to be uncomputable, what familiar functions (if any) are not effectively computable, and so forth. Other problems arise with respect to enumeration and decision procedures for sets of words or integers.

As stated this purpose is too indeterminate to be suitable for mathematical analysis, since the term "effective" has not been precisely defined (although informal arguments can be presented, as in Section 1 of this chapter). A specific, concrete model of the concept of "effective process" was introduced by A. M. Turing in 1936 [T1], in the form of a simple, idealized computing device, as described in Section II.2. Related work was done around the same time, by Church [C1], Post [P1], and others who proposed alternate models of this concept. These various formulations were all shown to be mutually equivalent, which led to the formulation of the famous Church–Turing thesis.

Church–Turing Thesis

The Turing machine is an accurate formalization of the intuitive concept of "effective process." Thus any computation done by a Turing machine is intuitively effective; conversely, any intuitively effective process can be simulated by a Turing machine. In particular.

   (i)   a function is effectively computable if and only if it is Turing computable;
   (ii) a set or predicate is effectively decidable if and only if it is Turing decidable (recursive);
   (iii) a set or predicate is effectively enumerable if and only if it is recursively enumerable.

4.2   Significance

Adoption of this thesis as a working assumption has several interesting consequences. As one example, it is quite common in the literature of computability (or recursive function) theory to sketch an informal algorithm to compute a function, and then appeal to the Church–Turing thesis to show that there is a Turing machine which will compute the function.

As another consequence, the Church–Turing thesis makes a formal theory of computability possible, since it implies that a clear-cut boundary can be established between the class of all computable functions and the class of uncomputable functions.

For example, in practice a statement such as "function f is uncomputable," or "problem P is unsolvable" often means merely that the natural approach to computing the function or solving the problem does not work. However the statement "f is not Turing computable" has a much stronger meaning. It asserts that there is *no Turing machine* which computes f, regardless of its computational method. Such a statement is extremely strong, since it says (by the Church–Turing thesis) not only that currently envisioned approaches must fail, but that no such approach can ever be envisioned which will compute f and be effective at the same time.

To see exactly what this means, let φ be a function of one argument which is not Turing computable, for example the function φ of Section II.3. It may in fact be possible to show that $\varphi(0) = 0$, $\varphi(1) = 1$, $\varphi(2) = 2$, and $\varphi(3) = 4$, up to some small finite limit, by a careful analysis of the possible behavior of all possible Turing machines with the corresponding numbers of instructions. In fact this has been done, by Rado and others as the "Busy Beaver" problem [R1].

However the statement "φ is not Turing computable" means that there can be no *general* effective method which, when given any arbitrary n, will eventually halt and produce $\varphi(n)$.

An interesting analogy can be drawn with the classically unsolvable problems of trisection of an angle, or solution of the fifth degree equation. For each problem we are given a well-defined set of acceptable tools, or methods which can be applied. For example the tools which can be used for the angle trisection problem are an unmarked ruler and a compass; the data is the angle between a given pair of straight lines. For the equation problem, the given data is the coefficients of a fifth-degree polynomial in one variable; an admissible method is a finite expression in these coefficients, built up by use of the operations signified by: $+$, $-$, $*$, $/$, and the taking of nth roots, for $n = 1, 2, 3, \ldots$. In each case, proofs have been constructed which show that these sets of tools are inadequate for their problems.

The situation for Turing machines is exactly similar: The problem to be solved is the general computation of a function φ, the tools available are Turing machine programs, and the proof shows that no Turing machine can compute φ. However there is an additional factor, the Church–Turing thesis.

According to this, the proof that no Turing machine can compute φ implies in addition that there can be no other effective process *of any kind* which computes φ. Stated in the contrapositive, if any method is devised which will compute φ, then this method must involve at least one step or operation which is not effective.

We now sketch three of the main arguments in favor of the Church–Turing thesis.

First, our classes of functions and predicates have very strong closure properties. If functions $f_1, f_2, \ldots$ are Turing computable and a function g is obtained from them by any of a large variety of natural effective finitary operations, then g will also be Turing computable. This means that it would be difficult to exhibit an effective function which is not Turing computable, because to do so would be necessary to define it in terms of operations which were finitary, but could not be reduced to those operations which have been shown to preserve Turing computability.

Second, there has been a variety of extremely diverse attempts to define and characterize the class of all effective processes. These include the Turing machine, the general recursive functions and the μ-recursive functions of Gödel, Herbrand, and Kleene, Church's idea of λ-calculability, Post's production systems, the existentially definable predicates of Smullyan, the Markov algorithm, and various types of register machines intended to simulate modern digital computers. *Without exception*, each of these formulations has been shown to be equivalent to all the others. These different formulations have very little similarity. The fact that such varied formulations all define the same class of computable functions, is a very strong point for the generality of any one of them.

The third argument is that the classes we study contain a *universal* effective process which is capable of simulating any other process in the class, or even itself. This is the universal Turing machine, which we will construct. Its existence corresponds to the idea of a "general-purpose digital computer" which is capable of simulating any other computer. In fact this concept of simulation seems to be quite central to the theory, since all of the proofs of equivalence of various formulations involve at some point a technique of using one system to simulate another.

4.3   Plan

It is of course impossible to present a formal proof of the Church–Turing thesis, since it concerns the relation between a formal system and an intuitive concept. However it is possible to give evidence in its favor, of the kinds mentioned in the previous subsection. It is possible to view the entire purpose of this book in that context.

Chapters III and IV contain a detailed proof of the equivalence of effective processes via Turing machines, with effectiveness defined by means of logical formulas (that is, the S-rudimentary and existentially definable predicates). Chapter III shows that the operation of Turing machines can be simulated by S-rudimentary formulas, using the elegant syntactic methods due to Smullyan [S1]. The machinery set up in this chapter will also be of use in the remainder of the book, particularly in Chapter VI.

Chapter IV shows in turn that Turing machines can be used to evaluate S-rudimentary and existentially definable formulas, thus establishing equivalence of these formulations.

Numerous consequences follow from this equivalence, mostly contained in Chapter V. The Normal Form Theorem (originally due to Kleene [K1]) follows directly, showing that the class of all recursive partial functions possesses a particularly simple, uniform characterization. Other results in Chapter V include a variety of positive and negative closure properties, including a formal proof that not every recursively enumerable set is recursive, and that the halting problem and some other problems are not effectively decidable.

The generality of the approach of Chapter III becomes evident in Chapter VI, in which the same methods are used to show the equivalence of Turing computability to three other formulations which have received much attention: the equational calculus of Kleene, and the semi-Thue and general canonical production systems of Post.

III DESCRIPTION OF TURING MACHINES BY PREDICATES

In this chapter we define the S-rudimentary predicates and establish a number of their properties which are useful in the description of Turing machines and their computations. We then describe an encoding scheme which assigns to every Turing machine, instantaneous description, and computation a unique word called a *Gödel word*. The actions of the Turing machine can be described completely and precisely by means of predicates involving these Gödel words. We show that these descriptive predicates are all S-rudimentary.

This implies immediately that every recursively enumerable predicate is existentially definable and that every recursive function is μ-rudimentary. The converses of these two theorems will be shown in the next chapter, thus establishing the equivalence of two very different formulations of computability

More generally, the machinery which we set up in this chapter (particularly that involving the encoding of sequences) will be useful in Chapter VI, in which it is used to show that several other formulations of computability are equivalent to Turing computability.

1. S-RUDIMENTARY PREDICATES

Suppose x and y are words over an alphabet A. We say that x is **part** of y if x is a subword of y; that is, if $y = uxv$ for some words u, $v \in A^*$. We write this as "$x\mathbf{P}y$." We say that x **begins** y, or that x **ends** y if $y = xu$ or $y = ux$, respectively, for some word $u \in A^*$. These are written as "$x\mathbf{B}y$" and "$x\mathbf{E}y$." Note that $u = v = \lambda$ is allowed, so $x\mathbf{P}x$, $x\mathbf{B}x$, and $x\mathbf{E}x$ are always true.

We now define the class of S-rudimentary predicates to be the smallest class of predicates containing $xy = z$ and closed under $\wedge$, $\vee$, $\neg$, explicit transformation, and quantification over subwords. This class is due originally to Smullyan [S1], and was used by him to prove equivalence of the existentially definable predicates with the predicates definable by elementary formal systems (yet another version of computability). The reader may wish to review Section I.3.

DEFINITION III.1.1

The class of **S-rudimentary** predicates is defined as follows:

(i)   The ternary predicate $xy = z$ is S-rudimentary.
(ii) Any explicit transform of an S-rudimentary predicate is S-rudimentary.
(iii) If $R(\vec{x}_n)$ and $S(\vec{x}_n)$ are S-rudimentary, then so are $\neg R(\vec{x}_n)$, $R(\vec{x}_n) \wedge S(\vec{x}_n)$, and $R(\vec{x}_n) \vee S(\vec{x}_n)$.
(iv) If $R(\vec{x}_n, y)$ is S-rudimentary, then so are $\exists z(z\mathbf{P}y \wedge R(\vec{x}_n, z))$ and $\forall z(z\mathbf{P}y \Rightarrow R(\vec{x}_n, z))$. We abbreviate these by $(\exists z)_{\mathbf{P}y} R(\vec{x}_n, z)$ and $(\forall z)_{\mathbf{P}y} R(\vec{x}_n, z)$, respectively.
(v)   No predicate is S-rudimentary unless it can be shown to be so by Rules (i) through (iv).   ∎

Throughout this chapter A will denote an alphabet containing at least two symbols, denoted by the digits 1 and 2. A **tally** is defined to be any word of the form 1^k, where $k \geq 0$.

THEOREM III.1.2

The following predicates are all S-rudimentary:

1. $x = y$, $x\mathbf{P}y$, $x\mathbf{B}y$, $x\mathbf{E}y$.
2. $x_1 x_2 \cdots x_n = y$, $x_1 \cdots x_n \mathbf{P}y$, $x_1 \cdots x_n \mathbf{B}y$, $x_1 \cdots x_n \mathbf{E}y$, for each value of $n = 1, 2, 3, \ldots$.
3. Any predicate $R(\vec{x}_n)$ which is true for only finitely many $\vec{x}_n$.
4. $|x| > n$, for each value of $n = 0, 1, 2, 3, \ldots$.
5. Tally(x) (true iff $x = 1^k$ for some $k \geq 0$).

PROOF

We show each of these S-rudimentary by exhibiting an equivalent formula which is built up from the relation $xy = z$ by the S-rudimentary operations:

1. $x = y \Leftrightarrow x\lambda = y$ (an explicit transform of $xy = z$);
    $x\mathbf{P}y \Leftrightarrow (\exists z)_{\mathbf{P}_y} z = x$; $\quad x\mathbf{B}y \Leftrightarrow (\exists u)_{\mathbf{P}_y} y = xu$; $\quad$ and
    $x\mathbf{E}y \Leftrightarrow (\exists u)_{\mathbf{P}_y} y = ux$.

2. $x_1 x_2 \cdots x_n = y$
    $\Leftrightarrow (\exists z_1, z_2, \ldots, z_n)_{\mathbf{P}_y}(z_1 = x_1 \wedge z_2 = z_1 x_2 \wedge z_3 = z_2 x_3$
    $\wedge \cdots \wedge z_n = z_{n-1} x_n \wedge z_n = y)$;
    $x_1 \cdots x_n \mathbf{P}y \Leftrightarrow (\exists z)_{\mathbf{P}_y} z = x_1 \cdots x_n$;
    $x_1 \cdots x_n \mathbf{B}y \Leftrightarrow (\exists z)_{\mathbf{P}_y}(z\mathbf{B}y \wedge z = x_1 \cdots x_n)$;
    and
    $x_1 \cdots x_n \mathbf{E}y \Leftrightarrow (\exists z)_{\mathbf{P}_y}(z\mathbf{E}y \wedge z = x_1 \cdots x_n)$.

3. Let the n-tuples for which $R(\vec{x}_n)$ is true be
    $$(r_1^1, \ldots, r_n^1), (r_1^2, \ldots, r_n^2), \ldots, (r_1^k, \ldots, r_n^k).$$
    Then R is S-rudimentary be the following:
    $$R(\vec{x}_n) \Leftrightarrow (x_1 = r_1^1 \wedge x_2 = r_2^1 \wedge \cdots \wedge x_n = r_n^1)$$
    $$\vee (x_1 = r_1^2 \wedge \cdots \wedge x_n = r_n^2)$$
    $$\vee \cdots \vee (x_1 = r_1^k \wedge \cdots \wedge x_n = r_n^k).$$

4. $|x| \leq n$ is true for finitely many x, and $|x| > n \Leftrightarrow \neg |x| \leq n$. Thus $|x| > n$ is also S-rudimentary.
5. Tally$(x) \Leftrightarrow (\forall s)_{\mathbf{P}_x}(s = \lambda \vee 1\mathbf{B}s)$. $\quad\blacksquare$

These will be used extensively later in this chapter.

Exercises

Let $A = \{1, 2\}$ in the following:

1. Define $R(x)$ to be true iff every occurrence of the symbol "1" in x is immediately followed by a "2." A natural first attempt to show that R is rudimentary would be to display the following:

$$R(x) \Leftrightarrow (\forall u)_{\mathbf{P}x}(u = 1 \Rightarrow u2\mathbf{P}x).$$

Unfortunately this does not define the desired predicate R.

(a) What is wrong with this definition?
(b) Construct a formula which defines R correctly.

2. Show that the 4ary predicate $uv = xy$ is S-rudimentary. Using this, give a new and shorter proof that Tally(r) is S-rudimentary.
3. Suppose $R(x)$ is true iff x is of the form $21^{n_1}21^{n_2}2 \cdots 21^{n_k}2$, where $k \geq 1$ and each $n_j \geq 2$. Show that R is S-rudimentary.

2. TURING MACHINES—FORMAL DEFINITIONS

In this section we give formal definitions of several of the terms already defined informally. This redefinition is necessary, not because the previous definitions are inadequate for common mathematical use, but rather because the Turing machine itself is to be an object of mathematical study. This means that the Turing machine must be defined as a mathematical object whose interpretation is independent of intuitive considerations. The student is urged to go through very carefully, and to verify for himself that the formal definitions are merely restatements of the earlier less formal definitions. A detailed understanding of these formal definitions is absolutely necessary for a real comprehension of the constructions and theorems in the remaining sections of Chapter III.

DEFINITION III.2.1

A **Turing machine** is a pair $Z = (A, P)$, where A is an alphabet not containing the symbols $*$ or : and P is a word of the form $p_1 : i_1$

$p_2 : i_2 \quad p_3 : \cdots p_k : i_k \quad p_{k+1}$, subject to the conditions below. Any word of this form which satisfies these conditions is called a **program**.

> (i) $p_1, p_2, \ldots, p_{k+1}$ are distinct symbols not in A, and are called **instruction labels**. The *initial label* of P is p_1, and the *final label* is p_{k+1}.
>
> (ii) Each $i_j (1 \leq j \leq k)$ is of one of the following forms, where $R, L, J \notin A \cup \{*, p_1, p_2, \ldots, p_{k+1}\}$; each i_j is called an **instruction**:
>
> > (a) R,
> > (b) L,
> > (c) a, where $a \in A \cup \{*\}$,
> > (d) Jap_l, where $a \in A \cup \{*\}$ and $1 \leq l \leq k + 1$.   ∎

This is the formal definition of a Turing machine. Note that we have simplified the informal definition slightly by omitting the instruction "*Jq*." This is no loss of generality, since the sequence "$p : Jq$" can be replaced by the equivalent sequence $p : J*q \quad p_1 : Ja_1q \quad p_2 : Ja_2q \cdots$ $p_m : Ja_mq$, where $A = \{a_1, a_2, \ldots, a_m\}$.

In order to describe computations formally, we next define an *instantaneous description* of a Turing machine Z. This is a word which is intuitively a "snapshot" of the total state of Z at a single instant. An instantaneous description (or **I.D.**, for short) must thus contain three items of information: the current contents of the tape of Z; the position of the scanning head on that tape; and the label of the instruction to be executed next. All this can be specified by a word of the form *upav* where a is the scanned symbol, p is the label, and *uav* is a string over $A \cup \{*\}$ which represents a contiguous portion of the tape which contains all nonblank symbols, as well as the scanned square.

We shall consider as identical any two I.D.s *upv* and *u'pv'* which denote the same infinite tape; the only differences which can occur are in the number of *s at the start of u and u', or at the end of v and v'.

DEFINITION III.2.2

Let $Z = (A, P)$ as above. An **instantaneous description** of Z is any word α of the form $\alpha = upav$, where $u, v \in (A \cup \{*\})^*$, $a \in A \cup \{*\}$, and p is an instruction label. The symbol scanned in α is the a which follows p.

The **initial I.D. for** $x_1, \ldots, x_n \in (A \cup \{*\})^*$ is $p_1 x_1 * x_2 \cdots * x_n *$, where p_1 is the label of the first instruction of P. An I.D. $\alpha = upav$ is **final** iff p is the last instruction label in P.   ∎

If α and β are I.D.s, we write $\alpha \vdash_Z \beta$, or say "α **yields** β **immediately by** Z" if $\alpha = upav$ and β is the I.D. which results from α by applying the instruction in P whose label is p. A formal definition is given next.

DEFINITION III.2.3

Let $Z = (A, P)$ be a Turing machine as in Subsection III.2.1, and let α and β be instantaneous descriptions of Z. In the following, u, v, and $\bar{u}$ denote arbitrary strings over $A \cup \{*\}$, a, b, and c denote symbols from $A \cup \{*\}$, p_j and p_l denote instruction labels, and i denotes an instruction.

By definition, $\alpha \vdash_Z \beta$ if and only if α is of the form $\alpha = up_j av$, P contains $p_j : i$, and one of the following cases applies:

1. $i \in A \cup \{*\}$ and $\beta = up_{j+1} iv$; or
2. $i = Jcp_l$ and either
   (a) $a = c$ and $\beta = up_l av$, or
   (b) $a \neq c$ and $\beta = up_{j+1} av$; or
3. $i = R$ and either:
   (a) $\beta = uap_{j+1} v$ and $v \neq \lambda$, or
   (b) $\beta = uap_{j+1} *$ and $v = \lambda$; or
4. $i = L$ and either:
   (a) $\beta = \bar{u} p_{j+1} bav$ and $u = \bar{u} b$ for some $b \in A \cup \{*\}$; or
   (b) $\beta = p_{j+1} * av$ and $u = \lambda$.   ∎

Note that cases 3 and 4 above correctly account for the addition of $*$s which may result from a move left or right.

DEFINITION III.2.4

Let $Z = (A, P)$ be a Turing machine as in Definition III.2.1, and let α, β be instantaneous descriptions of Z. We say "α **yields** β **by** Z," written "$\alpha \vdash_Z^* \beta$" if and only if there is a sequence $\alpha_1, \alpha_2, \ldots, \alpha_l$ of instantaneous descriptions with $l \geq 1$ such that

(i) $\alpha = \alpha_1$ and $\beta = \alpha_l$
(ii) $\alpha_i \vdash_Z \alpha_{i+1}$ for $i = 1, 2, \ldots, l - 1$.

ITHACA COLLEGE LIBRAR
WITHDRAWN

Note that $\alpha = \alpha_1 = \beta$ in case $l = 1$; thus $\alpha \vdash_Z^* \alpha$ is true for any I.D. α.

A **computation** is any sequence $\alpha_1, \alpha_2, \ldots, \alpha_l$ such that $\alpha_i \vdash_Z \alpha_{i+1}$ for $i = 1, 2, \ldots, l - 1$. Z **halts** when given α iff there is a computation beginning with α such that its last I.D. is final; if there is no such computation, then Z **loops** with α.   ∎

If Z is understood from context, we will write $\alpha \vdash \beta$ and $\alpha \vdash^* \beta$ in place of $\alpha \vdash_Z \beta$ and $\alpha \vdash_Z^* \beta$.

Exercises

1. If $p : a \quad q$ is part of program P, does P necessarily have an instruction with label q?
2. Can $p : a \quad p$ be part of a program P?
3. Under what circumstances can $\alpha \vdash_Z \alpha$ be true? When can $\alpha \vdash_Z^* \alpha$ be true?
4. Give a necessary and sufficient condition on an I.D. α so that $\alpha \vdash \beta$ is true for no I.D. β.
5. Give a formal definition of one of the machines in Section II.2 [merely give a pair $Z = (A, P)$].
6. Give an example of a computation by one of these machines.
7. List an example of each case of 3 and 4 of Definition III.2.3.
8. Prove the remark which follows Definition III.2.3.
9. Give definitions of the terms "recursive function," "recursively enumerable set" and "recursive set," using the present terminology.
10. Show that a predicate is recursively enumerable if it is recursive. *Hint:* The proof should take the following form. "Suppose $R(\vec{x}_n)$ is recursive. Then there is a Turing machine $Z = (A, P)$ as in Definition III.2.1. If we modify Z as follows, we obtain a new Turing machine $Z' = (A, P')$ which we claim has R as its domain." At this point, give the definition of Z' and prove that Z' behaves as required.
11. Show that $\neg R$ is recursive if R is recursive.

3. GÖDEL WORDS AND THE BASIC SIMULATION PREDICATES

Our basic idea is to encode the various Turing machine constructs as words over the alphabet A, and to define predicates which describe in encoded form the interrelations among programs, inputs, instantaneous descriptions, and computations. This general approach was first devised by Kurt Gödel [G1] (for a completely different type of system) so we call these encoded programs, *Gödel words*. Two conditions must be satisfied for an assignment of Gödel words to programs, instantaneous descriptions and computations:

(i) The Gödel word $gw(x)$ corresponding to an object x must be effectively obtainable from x, and conversely; and

(ii) Different objects must have different Gödel words; that is, $x \neq y$ implies $gw(x) \neq gw(y)$.

Leaving the exact specification of $gw(x)$ undefined for the moment we now define the main predicates used in Turing machine simulation. Our only assumptions are that $gw(x)$ is always a word over A (A is fixed), and that (i) and (ii) above hold.

DEFINITION III.3.1

The **basic simulation predicates** are defined as follows:

1. TM(z) is true if $z = gw(Z)$ for some Turing machine $Z = (A, P)$.

2. Yield(z, α, β) is true iff there is a Turing machine Z with instantaneous descriptions α', β' such that $\alpha' \vdash_Z \beta'$, $z = gw(Z)$ and α and β are Gödel words of α' and β', respectively.

3. Comp(z, y) is true iff there is a Turing machine Z with computation $\alpha_1, \alpha_2, \ldots, \alpha_l$ such that $z = gw(Z)$ and y is the Gödel word of the computation.

4. $T_n(z, x_1, \ldots, x_n, y)$ is true iff Comp(z, y) is true as in 3, and α_1 is the initial I.D. for $x_1, \ldots, x_n$, and α_l is final.

5. The function, $U : A^* \to A^*$ is defined by: $U(y) = w$ if y is an encoded computation as in 3, and w is the content of α_l (defined as in Section II.2 "Computation of Functions"). ∎

From these definitions we immediately obtain the following. Recall the definition of the minimalization operator μ at the end of Section I.3.

THEOREM III.3.2

(i) If $P(\vec{x}_n)$ is a recursively enumerable predicate over A, there is a word $z_0 \in A^*$ such that, for all $x_1, \ldots, x_n \in A^*$

$$P(\vec{x}_n) \Leftrightarrow \exists y \, T_n(z_0, \vec{x}_n, y). \tag{1}$$

(ii)   If $f(\vec{x}_n)$ is a recursive function over A, there is a word $z_0 \in A^*$
such that, for all $x_1, \ldots, x_n \in A^*$, either

$$f(\vec{x}_n) = U(\mu y T_n(z_0, \vec{x}_n, y)), \tag{2}$$

or both sides of this equation are undefined.

PROOF

If P is recursively enumerable there must be a Turing machine Z
which accepts it. Let $z_0 = gw(Z)$. By definition, $P(\vec{x}_n)$ is true if Z
eventually halts when started with $x_1, \ldots, x_n$, that is, iff there is a
computation $\alpha_1, \alpha_2, \ldots, \alpha_l$ beginning with the initial I.D. for $x_1, \ldots, x_n$,
such that α_l is final. Thus $P(\vec{x}_n)$ is true iff for some y, $T_n(z_0, \vec{x}_n, y)$ is
true.

Now suppose that f is computed by Turing machine Z, and let
$z_0 = gw(Z)$. For each $x_1, \ldots, x_n \in A^*$, $f(\vec{x}_n)$ is defined iff Z eventually
halts when started with $p_1 x_1 * \cdots * x_n$, that is, iff $\mu y T_n(z_0, \vec{x}_n, y)$ is
defined. Thus the two sides of Equation (2) are either both defined or
both undefined. Now, if $f(\vec{x}_n)$ is defined, then Z has a computation
$\alpha_1, \alpha_2, \ldots, \alpha_l$ such that $\alpha_1 = p_1 x_1 * \cdots * x_n$ and the content of α_l is
$f(\vec{x}_n)$. Thus $\mu y T_n(z_0, \vec{x}_n, y)$ is defined and $U(y) = f(\vec{x}_n)$ for this y, so
that Equation (2) is true.   ∎

This theorem is closely related to the important normal form
theorems to be proved in Chapter V. We will first define a system of
assigning Gödel words to Turing machines, instantaneous descriptions, and computations, and then prove that the corresponding U,
T_n, and so forth, are all S-rudimentary. This immediately implies
by the above lemma that every recursively enumerable predicate is
existentially definable, and that every recursive function is μ-rudi-
mentary.

The Turing Machine Code

DEFINITION III.3.3

Let $Z = (A, P)$ be a Turing machine with $P = p_1 : i_1 \quad p_2 : i_2 \cdots$
$p_k : i_k \quad p_{k+1}$. Let $gw(Z)$ be the word obtained from P by replacing every
symbol s in P by $gw(s)$ according to the following rules:

$$
\begin{aligned}
gw(a) &= a & &\text{if } a \in A, \\
gw(*) &= 212 & &\text{(abbreviated } *\text{)}, \\
gw(R) &= 21^2 2 & &\text{(abbreviated } \mathbf{R}\text{)}, \\
gw(L) &= 21^3 2 & &\text{(abbreviated } \mathbf{L}\text{)}, \\
gw(J) &= 21^4 2 & &\text{(abbreviated } \mathbf{J}\text{)}, \\
gw(:) &= 21^5 2 & &\text{(abbreviated } :\text{)}, \\
gw(p_i) &= 21^{5+i} 2 & &\text{for } i = 1, 2, \ldots \quad \blacksquare
\end{aligned}
$$

Notice that Z can be completely reconstructed from $z = gw(Z)$, except for a possible renaming of $p_1, \ldots, p_{k+1}$. This is true only because of the fact that P may not contain two or more symbols of A in adjacent positions. Ambiguity can result if this code is applied to entities which are not programs. For example, $gw(* 2112 *) = gw(* R *) = gw (212R212) = \cdots = gw(2122112212)$.

A Technique for Encoding Sequences

A single, fixed code such as the one just given for Turing machines is possible because of the fact that Turing machine programs have a restricted format. A fixed code is more difficult for I.D.s, computations, and so forth, because these may contain any combination whatever of 1s and 2s. In order to avoid this difficulty we use a technique due to Quine [Q1], called a "longest-tally code."

Given a sequence $\alpha_1, \ldots, \alpha_k$ ($k \geq 1$) of words over an alphabet A which contains two symbols "1" and "2," let $\Delta = 2r2$ be a string such that r is a tally, and no tally which occurs among $\alpha_1, \ldots, \alpha_k$ is as long as r. The sequence is encoded as $\Delta\alpha_1 \Delta\alpha_2 \cdots \Delta\alpha_k$. This sequence can be easily (and uniquely) broken up into its original components, by finding the Δ at its beginning, and then isolating all Δs. If each Δ except the first is replaced by a comma, we have the sequence $\alpha_1, \ldots, \alpha_k$ back in its original form. We shall call such a string Δ a *separator*.

Note that each α_i may itself be an encoded sequence, for example, α_1 might equal $\#\beta_1\#\beta_2 \cdots \#\beta_k$, where $\#$ is another separator. Clearly the tally in Δ must be longer than the tally in $\#$ in this case.

For example, let $\alpha_1 = 1211$, $\alpha_2 = 12$, and $\alpha_3 = 2112$. An admissible value Δ would be $21^4 2$ ($21^3 2$ would also work), and the encoded form would be $21^4 2121121^4 21221^4 22112$.

This technique will also be used in Chapter VI, to aid in simulation of other types of computing devices.

Encoding Instantaneous Descriptions

DEFINITION III.3.4

Let $\alpha = up_i v$ be an instantaneous description, and let $\sharp = 2r12$, where r is a tally longer than any tally among u, $gw(p_i)$, and v. By definition

$$gw^{\sharp}(\alpha) = \sharp u' \sharp gw(p_i) \sharp v'$$

where u' and v' are obtained from u and v, respectively, by replacing each occurrence of the symbol " $*$ " by $2r2$. ∎

For example if $\alpha = 2*1p_3 22$, then an admissible value of r would be 1^9, in which case

$$gw^{\sharp}(\alpha) = 21^{10}2221^921 21^{10}221^8221^{10}222.$$

Clearly $gw^{\sharp}(\alpha)$ is an encoded sequence, so $\sharp$, u', $gw(p_i)$, and v' can be uniquely obtained from it. Further, p_i is uniquely determined by $gw(p_i)$, and if we let $\sharp = 2r12$, then u and v may be uniquely obtained from u' and v' by replacing each occurrence of $2r2$ by " $*$."

Computations

DEFINITION III.3.5

Let $Z = (A, P)$ be a Turing machine and let $\alpha_1, \alpha_2, \ldots, \alpha_k$ be a computation by Z. Let $\sharp = 2r12$ be the shortest word such that $gw^{\sharp}(\alpha_1)$, $\ldots$, $gw^{\sharp}(\alpha_k)$ are all defined.

The Gödel word of the computation $\alpha_1, \alpha_2, \ldots, \alpha_k$ is defined to be $\Delta \alpha_1' \Delta \alpha_2' \Delta \cdots \Delta \alpha_k'$, where $\Delta = 2r112$ and $\alpha_1' = gw^{\sharp}(\alpha_1), \ldots, \alpha_k' = gw^{\sharp}(\alpha_k)$. ∎

For example, consider the two-step computation α_1, α_2 with $\alpha_1 = 2p_1*$ and $\alpha_2 = 2p_2 1$. Then $*$ is coded as 21^82 and $\sharp = 21^92$, so $gw^{\sharp}(\alpha_1) = 21^92221^9221^6221^9221^82$ and $gw^{\sharp}(\alpha_2) = 21^92221^9221^7221^921$. This implies that $\Delta = 21^{10}2$, and that the Gödel word of α_1, α_2 is $\Delta\alpha'_1\Delta\alpha'_2$, that is,

$$21^{10}221^9 2221^9 221^6 221^9 221^8 221^{10} 221^9 2221^9 221^7 221^9 21$$

or

$$\Delta\#2\#gw(p_1)\#s\Delta2\#gw(p_2)\#1,$$

where $s = 21^8 2$.

It is easily verified that if y is the Gödel word of a computation, then y completely determines $\alpha_1, \alpha_2, \ldots, \alpha_k$.

Exercises

1. Find the Turing machine Z whose Gödel word $z = gw(Z)$ is
    $21^6 221^5 2121^7 221^5 221^4 2121^6 221^8 2$.
2. Find the instantaneous description α and the word $\#$ for which
    $y = gw^\#(\alpha)$, where $y = 21^9 2221^8 2221^9 221^7 221^9 2121^8 21$.
3. Write formulas using the basic simulation predicates which
    express the statements "Z halts for every input pair (x, y)"
    and "whenever Z halts, its tape contains 121."

4. THE BASIC SIMULATION PREDICATES
ARE S-RUDIMENTARY

We now show that these predicates are S-rudimentary. First, we define some auxiliary predicates which are useful for description of Turing machines.

LEMMA III.4.1

The following predicates are all S-rudimentary:

1. Label (p),   true iff $p = gw(p_i)$ for some label p_i.
2. INSW(i),   true iff $i = *$ or $i \in A$.
3. INSR(i),   true iff $i = \mathbf{R}$.
4. INSL(i),   true iff $i = \mathbf{L}$.
5. INSJ(i, c, q),   true iff $i = gw(Jc'q')$, where $c \in A \cup \{*\}$, $c = gw(c')$, $q = gw(q')$, and q is a label.
6. Inst(z, i, p),   true iff $\mathbf{p} : \mathbf{i}$ occurs in z and i is the Gödel word of an instruction which, if a Jump, refers to a label whose coded form is a subword of z.

7. Next(p, q),   true iff for some $i \geq 1$, $p = gw(p_i)$ and $q = gw(p_{i+1})$.

8. TM(z),   true iff $z = gw(Z)$ for some Turing machine Z.

PROOF

Numbers 2, 3, and 4 are finite predicates, and so are S-rudimentary by Lemma III.1.2. The remaining are S-rudimentary by the following formulas:

1.   Label(p) $\Leftrightarrow (\exists r)_{\mathbf{P}p}$(Tally($r$) $\wedge p = 2r2 \wedge |r| > 5$),
5.   INSJ(i, c, q) $\Leftrightarrow$ Label(q) $\wedge i = $ J$cq \wedge$ INSW(c),
6.   Inst(z, i, p) $\Leftrightarrow$ {INSR(i) $\vee$ INSL(i) $\vee$ INSW(i)
                  $\vee (\exists c, q)_{\mathbf{P}i} [$INSJ($i$, c, q) $\wedge (q : \mathbf{P}z \vee q\mathbf{E}z)]$}
              $\wedge$ Label(p) $\wedge p : i\mathbf{P}z$,
7.   Next(p, q) $\Leftrightarrow$ Label(p) $\wedge (\exists r)_{\mathbf{P}p}(p = 2r2 \wedge q = 2r12)$,
8.   TM(z) $\Leftrightarrow 21^62 : \mathbf{B}z \wedge (\forall u, p)_{\mathbf{P}z} \{up : \mathbf{B}z \wedge$ Label(p)
                 $\Rightarrow (\exists q, i)_{\mathbf{P}z} [$Next($p$, q) $\wedge$ Inst(z, i, p)
                       $\wedge (up : iq : \mathbf{B}z \vee z = up : iq)]$}.

To verify this last formula first suppose that TM(z) is true, so $z = gw(Z)$ for some Turing machine $Z = (A, P)$. Then "p_1:" begins P, so 21^62: must begin z. Further, any occurrence of a label "p_j:" in P must be followed by an instruction "i" and the label "p_{j+1}." If this label "p_{j+1}" is not the final label, it will be followed by a colon; otherwise it will end P. Thus the part of the formula in curly braces must be true for any u and p.

Conversely, suppose that the formula is true of a word z. Then z must be of the form $z = 21^62 : w_1$. If we let $u = \lambda$ and $p = 21^62$, we see by the bracketed clause that there is an instruction i_1 such that either $z = 21^62 : i_1 21^72$, or $z = 21^6 : i_1 21^72 : w_2$ for some word w_2. In the first case TM(z) is certainly true. In the second case, we can let $u = 21^62 : i_1$ and $p = 21^72$, and see similarly that either $z = 21^62 : i_1$ $21^72 : i_2$ 21^82, or $z = 21^62 : i_1$ $21^72 : i_2$ $21^82 : w_3$ for some word w_3 and instruction i_2. Continuing this way, we see that z must in all cases have the form

$$z = gw(p_1 : i_1 \quad p_2 : i_2 \cdots p_k : i_k \quad p_{k+1}),$$

so that TM(z) is true.   ∎

In preparation for $ID(\alpha)$, Final (z, α), and various other constructions, we define some predicates involving sequences and show that they are S-rudimentary.

LEMMA III.4.2

The following predicates are all S-rudimentary.

1. $Seq(y, \Delta)$,   true iff y is of the form $\Delta\alpha_1\Delta\alpha_2 \cdots \Delta\alpha_k$, where $k \geq 1$ and $\Delta = 2r2$, where r is a tally longer than the longest tally among $\alpha_1, \alpha_2, \ldots, \alpha_k$.
2. $First(y, \alpha)$ true iff y is as above and $\alpha = \alpha_1$.
3. $Last(y, \alpha)$,   true iff y is as above and $\alpha = \alpha_k$.
4. $Adj(\alpha, \beta, y)$,   true iff y is as above and for some $i(1 \leq i < k)$ $\alpha = \alpha_i$ and $\beta = \alpha_{i+1}$.

PROOF

1. $Seq(y, \Delta) \Leftrightarrow \Delta\mathbf{B}y \wedge (\exists r)_{\mathbf{P}y}\{tally(r) \wedge \Delta = 2r2$

$$\wedge \neg 2r2r2 \ \mathbf{P}y \wedge (\forall u, v)_{\mathbf{P}y}[urv = y \Rightarrow 2\mathbf{E}u \wedge 2\mathbf{B}v]\}.$$

If y is an encoded sequence it is immediate that the right-hand side of this formula is true. Conversely suppose the right-hand side is true. Then $\Delta = 2r2$, where r is a tally. The condition that $2r2r2$ is not part of y implies that two occurrences of Δ in y cannot overlap, so that y can be uniquely expressed in the form $\Delta\alpha_1\Delta\alpha_2 \Delta \cdots \Delta\alpha_k$, where the αs do not contain Δ as a subword. The condition $(\forall u, v)_{\mathbf{P}y}(urv = y \Rightarrow 2\mathbf{E}u \wedge 2\mathbf{B}v)$ implies that r can occur only as part of Δ, so that r is longer than the longest tally among $\alpha_1, \ldots, \alpha_k$.

The remaining predicates are more easily verified.

2. $First(y, \alpha) \Leftrightarrow (\exists\Delta)_{\mathbf{P}y}(Seq(y, \Delta) \wedge \neg \Delta\mathbf{P}\alpha \wedge (\Delta\alpha = y \vee \Delta\alpha\Delta\mathbf{B}y))$.
3. $Last(y, \alpha) \Leftrightarrow (\exists\Delta)_{\mathbf{P}y}(Seq(y, \Delta) \wedge \neg \Delta\mathbf{P}\alpha \wedge \Delta\alpha \ \mathbf{E}y)$.
4. $Adj(\alpha, \beta, y) \Leftrightarrow (\exists\Delta)\mathbf{P}_y(Seq(y, \Delta) \wedge \neg \Delta\mathbf{P}\alpha \wedge \neg \Delta \ \mathbf{P}\beta$
$$\wedge (\Delta\alpha\Delta\beta\mathbf{E}y \vee \Delta\alpha\Delta\beta\Delta \ \mathbf{P}y)). \blacksquare$$

We now proceed to define predicates which are useful for describing instantaneous descriptions, and initial and final I.D.s.

LEMMA III.4.3

The following predicates are S-rudimentary.

1.	Tape(u, r),	true iff u can be obtained from some word u' over $A \cup \{*\}$ by replacing each occurrence of $*$ by $2r2$, and r is a tally longer than any tally in u'.
2.	ID(α, $\sharp$),	true iff α is the encoded form, with $\sharp$ as separator, of an instantaneous description *upav*.
3.	Init(α, x_1, ..., x_n),	true iff $\alpha = gw^\sharp(\alpha')$ where α' is the initial I.D. for x_1, ..., x_n.
4.	Final(z, α),	true iff $\alpha = gw^\sharp(\alpha')$ where α' is a final I.D. for a Turing machine Z such that $z = gw(Z)$, and r is its separator.

1. Tape(u, r) $\Leftrightarrow$ Tally(r) $\wedge \neg\, 2r2r2Pu$
   $\wedge\ (\forall x, y)_{\mathbf{P}y}(u = xry \Rightarrow 2Ex \wedge 2By)$.
2. ID(α, $\sharp$) $\Leftrightarrow (\exists u, p, a, v, r)_{\mathbf{P}\alpha}[\alpha = \sharp u \sharp p \sharp av \wedge$ Seq(α, $\sharp$) $\wedge \sharp = 2r12$
   $\wedge$ Tape(u, r) $\wedge$ Label(p) $\wedge$ Tape(v, r)
   $\wedge\ (a \in A \vee a = 2r2)]$.

PROOF

   Note that this formula has a close correspondence to Definitions III.2.2 and III.3.4.

3. Init(α, $\vec{x}_n$) $\Leftrightarrow (\exists \sharp, r, s)_{\mathbf{P}\alpha}[$ID($\alpha$, $\sharp$) $\wedge \sharp = 2r12 \wedge s = 2r2$
   $\wedge\ \alpha = \sharp\sharp21^62\sharp x_1 s x_2 s \cdots x_n s]$.
4. Final(z, α) $\Leftrightarrow$ TM(z) $\wedge\ (\exists \sharp, p)_{\mathbf{P}\alpha}[$ID($\alpha$, $\sharp$)
   $\wedge$ Label(p) $\wedge pEz \wedge pP\alpha]$.   ∎

We now come to the heart of the simulation, by showing that Yield(z, α, β) is S-rudimentary.

LEMMA III.4.4

   Yield(z, α, β) is S-rudimentary.

PROOF

The following S-rudimentary formula defines Yield; note that its structure corresponds very closely to Definition III.2.3. A minor complication arises, since the symbol $*$ must be encoded differently in a Turing machine code than in an I.D. In particular, $*$ is coded as $*$ in a Turing machine, and as $2r2$ in an I.D. α for which $\# = 2r12$.

$\mathrm{Yield}(z, \alpha, \beta)$

$\Leftrightarrow \mathrm{TM}(z) \wedge (\exists p, q, p', i, c)_{\mathbf{P}z}(\exists \#, r, u, a, v, b)_{\mathbf{P}\alpha}(\exists \bar{u}, \tilde{v})_{\mathbf{P}\beta}$
$\{\mathrm{Inst}(z, i, p) \wedge \# = 2r12$
$\quad \wedge (a \in A \vee a = 2r2) \wedge (b \in A \vee b = 2r2)$
$\quad \wedge \mathrm{ID}(\alpha, \#) \wedge \alpha = \#u\#p\#av \wedge \mathrm{ID}(\beta, \#) \wedge \beta = \#\bar{u}\#q\#\tilde{v}$
$\quad \wedge \{[\mathrm{INSW}(i) \wedge \bar{u} = u \wedge \mathrm{Next}(p, q)$
$\qquad \wedge \{(i \neq * \wedge \tilde{v} = iv) \vee (i = * \wedge \tilde{v} = 2r2v)\}]$
$\quad \vee [\mathrm{INSJ}(i, c, p') \wedge \bar{u} = u \wedge \tilde{v} = av$
$\qquad \wedge \{(a = c \vee a = 2r2 \wedge c = *) \wedge q = p'$
$\qquad \vee \neg (a = c \vee a = 2r2 \wedge c = *) \wedge \mathrm{Next}(p, q)\}]$
$\quad \vee [\mathrm{INSR}(i) \wedge \bar{u} = ua \wedge \mathrm{Next}(p, q)$
$\qquad \wedge \{(v \neq \lambda \wedge \tilde{v} = v) \vee (v = \lambda \wedge \tilde{v} = 2r2)\}]$
$\quad \vee [\mathrm{INSL}(i) \wedge \mathrm{Next}(p, q)$
$\qquad \wedge \{(u = \bar{u}b \wedge \tilde{v} = bav)$
$\qquad \vee (u = \lambda \wedge \bar{u} = \lambda \wedge \tilde{v} = 2r2av)\}]\}\}\}.$ ∎

THEOREM III.4.5

The following predicates are S-rudimentary:

1. $\mathrm{Comp}(z, y)$,   true iff y is the encoded form of a sequence $\alpha_1, \alpha_2, \ldots, \alpha_l$ of instantaneous descriptions such that $\alpha_i \vdash \alpha_{i+1}$ for $1 \leq i < l$.

2. $T_n(z, x_1, \ldots, x_n, y)$,   true iff y is as above, and α_1 is the initial instantaneous description for $x_1, \ldots, x_n$ and α_l is final.

3. $U(y) = w$,   true if y is as in 1, and w is the content of α_l.

PROOF

1. $\mathrm{Comp}(z, y) \Leftrightarrow \mathrm{TM}(z) \wedge (\forall \alpha, \beta)_{\mathbf{P}y}[\mathrm{Adj}(\alpha, \beta, y) \to \mathrm{Yield}(z, \alpha, \beta)]$.
2. $T_n(z, x_1, \ldots, x_n, y)$
   $\Leftrightarrow \mathrm{Comp}(z, y) \wedge (\exists \alpha, \beta)_{\mathbf{P}y}[\mathrm{First}(y, \alpha) \wedge \mathrm{Init}(\alpha, x_1, \ldots, x_n)$
   $\wedge \mathrm{Last}(y, \beta) \wedge \mathrm{Final}(z, \beta)]$.
3. $U(y) = w \Leftrightarrow (\exists \alpha, \#, r, v)_{\mathbf{P}y}[\mathrm{Last}(\alpha, \alpha) \wedge \mathrm{Last}(\alpha, v) \wedge \mathrm{Seq}(\alpha, \#)$
   $\wedge \# = 2r12 \wedge \neg 2r2Pw \wedge (w = v \vee w2r2Bv)]$. ∎

5. EXISTENTIALLY DEFINABLE PREDICATES,
μ-RUDIMENTARY FUNCTIONS

In order to relate the results just obtained to Theorem III.3.2, we introduce some new terminology.

DEFINITION III.5.1

A predicate $R(\vec{x}_n)$ is **existentially definable** iff there is an S-rudimentary predicate $Q(\vec{x}_n, y)$ such that for all $x_1, \ldots, x_n \in A^*$

$$R(\vec{x}_n) \Leftrightarrow \exists y \, Q(\vec{x}_n, y).$$

A function $f(\vec{x}_n)$ is defined to be **S-rudimentary** iff

(a)   the predicate $f(\vec{x}_n) = y$ is S-rudimentary and
(b)   there is an index i $(1 \leq i \leq n)$ such that for all $\vec{x}_n$ such that $f(\vec{x}_n)$ is defined, $f(\vec{x}_n)$ is a subword of x_i.

A function $f(\vec{x}_n)$ is defined to be **μ-rudimentary** iff there is an S-rudimentary function $h(y)$ and predicate $Q(\vec{x}_n, y)$ such that for all $x_1, \ldots, x_n \in A^*$

$$f(\vec{x}_n) = h(\mu y \, Q(\vec{x}_n, y)). \blacksquare$$

THEOREM III.5.2

Suppose $\sharp A \geq 2$. Then every recursively enumerable predicate is existentially definable, and every recursive function is μ-rudimentary.

PROOF

The first part is immediate from Theorem III.3.2, part (i), and Theorem III.4.5. The second part follows from Theorem III.3.2, part (ii), Theorem III.4.5, and the observation that for every y, $U(y)$ is a subword of y (if defined). $\blacksquare$

In the next chapter we will prove the converses of these theorems, thus establishing equivalence of these concepts with Turing computability.

The most subtle part of this equivalence is the material just covered on the simulation of Turing machine activities by means of predicates. The S-rudimentary predicates were chosen for this task because they are the simplest class of predicates known to the author which are capable of describing Turing machine behavior in a natural manner. A more common choice is the primitive recursive functions and predicates used by Kleene [K1], Davis [D1], and others. In order to use those classes however, it is necessary to *arithmetize* Turing machines, computations, and so forth, that is, to encode them in terms of positive integers. This leads to even messier constructions, and necessitates the use of various devices from number theory which are perhaps less intuitive than our S-rudimentary encoding techniques.

IV DECISION OF PREDICATES BY TURING MACHINES

In this chapter we finish laying the foundations for the important theorems of Chapter V. In Chapter III it was shown that Turing machines can be described by S-rudimentary predicates; now we show that S-rudimentary predicates can be decided by Turing machines.

Section IV.1 contains tools and techniques for constructing complex Turing machines from simple ones. The ideas of "subroutines" and "flow charts," both terms borrowed from computer programming, are defined in this section, and are used extensively in Section IV.3. The use of subroutines makes it possible to use previously defined machines as elementary components of new machines. Flow charts are merely diagrammatic versions of programs, and are used mainly because the structure of our Turing algorithms is much more evident when displayed in two-dimensional form. In principle all our Turing machine constructions, including machines to decide the complex predicates U and T_n, could be carried out without the use of either subroutines or flow charts. In practice however this would be prohibitively tedious.

Section IV.2 contains the definition of standard computability, decidability, and acceptability.

The main results of Section IV.3 are the statements:

(i)   Every S-rudimentary predicate is recursive.

(ii) Every existentially definable predicate is recursively enumerable.

(iii) Every μ-rudimentary function is recursive.

One of the main results of Chapter III was that the converses of Statements (ii) and (iii) are also true, so that these classes of predicates and functions are identical. Since the classes are defined inductively, the proofs of Statements (i), (ii), and (iii) are also inductive. Consequently Section IV.3 is essentially a series of lemmas of the form "if P and Q are recursive (and so forth) and Θ is an operation, then $\Theta(P, Q)$ is recursive (and so forth)," where Θ can be any of the operations $\wedge$, $\vee$, $\exists$, minimalization, composition, and so forth.

Section IV.4 contains a proof that the use of auxiliary tape symbols does not increase the computational power of Turing machines. This result will be used in Chapter VI, but is not necessary for the normal form theorems and consequences of Chapter V.

In summary, this chapter shows a large number of closure properties of the Turing-defined classes of predicates and functions, and proves the second half of the equivalence between two rather divergent concepts: the recursively enumerable and the existentially definable predicates, and the recursive and the μ-rudimentary functions.

1. SUBROUTINES AND FLOW CHARTS

In this section we lay the foundations for some rather complex Turing machine constructions which will come later in this chapter. In order to simplify these constructions, we introduce two notational devices: subroutines and flow charts.

Subroutines

Certain sequences of instructions will with minor variations be common to many Turing machines. For example, in one machine's program we might find $11 : R$ $12 : Ja11$ $13 : Jb11$ and in another, $74 : R$ $75 : Ja74$ $76 : Jb$ 74. If $A = \{a, b\}$, these serve the same purpose: each causes the scanning head to search to the right, stopping at the first blank symbol $(*)$ it encounters. We shall give names to some of these sequences which are frequently used, and use the names as though they were actually instructions. For example, the sequence above is denoted by $R(*)$. A sequence of this sort is called a **subroutine**.

It should be clear that any Turing machine Z which uses subroutine names as instructions could be replaced by an entirely equivalent machine Z' which uses only a, R, L, and Jaq as instructions. Z' is obtained from Z by replacing each subroutine name by the corresponding sequence of instructions, and by renaming labels if necessary to avoid duplicate or undefined labels.

For example, let $R(*)$ denote the sequence $1 : R$ $2 : Ja1$ $3 : Jb1$. If $Z = (A, P)$ is a Turing machine using $R(*)$ as an instruction, with $A = \{a, *\}$ and

$$P = 1 : R(*)   2 : R   3 : Ja6   4 : R(*)   5 : R   6 : Ja2   7,$$

an equivalent basic Turing machine would be $Z' = (A, P')$, where

$$P' = 1 : R   1' : Ja1   1'' : Jb1   2 : R   3 : Ja6$$
$$4 : R   4' : Ja4   4'' : Jb4   5 : R   6 : Ja2   7$$

Flow Charts

A **flow chart** is a two-dimensional diagrammatic form of a program, in which labels are omitted as much as possible, (consistent with clarity) and arrows are used to indicate jumps. Our notation is similar to that of Hermes [H1]. We now give some basic rules governing the relation between flow charts and programs.

First, a flow chart is read from left to right, unless explicitly indicated otherwise by arrows. Labels may be and usually are omitted. For examples we have the flow charts following:

$$P_1 : aR*LLH \qquad \text{and} \qquad P_2 : aRR* \;\bullet.$$

These might correspond to the programs

$$P_1 = 1:a \quad 2:R \quad 3:* \quad 4:L \quad 5:L \quad 6$$

and

$$P_2 = 1:a \quad 2:R \quad 3:R \quad 4:* \quad 5:J1 \quad 6.$$

The symbol "H" (for "HALT") is used to indicate the end of a series of instructions, that is, the final label. This symbol will also be used in place of an unconditional jump to the final label, thus reducing the number of arrows.

A conditional jump (for example, "$Ja6$") is represented by one or two arrows, one labeled "a" or "$\neq a$" (or both may be labeled). A series of jumps may be indicated by appropriate labeling. For example, the program

$$P = 1:R \quad 2:Ja7 \quad 3:R \quad 4:Ja7 \quad 5:R \quad 6:Jb1 \quad 7$$

would appear in flow chart form as follows:

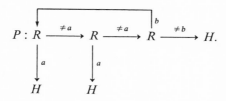

More complex conditional jumps may be represented by stacks of arrows. For example, suppose it is desired to replace the digits 1, 2, 3 by 2, 3, 1, respectively. A flow chart to accomplish this might be

If a single instruction or a large portion of a flow chart is repeated, say k times, we may abbreviate the chart by use of k as a superscript, enclosing the repeated portion in parentheses if necessary. Thus the following charts are equivalent:

$$LLRRRa \quad \text{and} \quad L^2 R^3 a;$$

$$LaLaLaR \quad \text{and} \quad (La)^3 R.$$

A superscript of 0 shall be taken to mean that the machine does not occur at all. For example, the chart $L(R(*)R)^n *$ stands for $L *, L R(*) R *,$ $L R(*) R R(*) R *, \ldots$ as n assumes the values 0, 1, 2, $\ldots$.

A Library of Basic Turing Machines

We now define a collection of basic Turing machines which will be very frequently used as subroutines in Turing machine constructions. They will be defined by means of the flow charts just described. Of course, any of these flow charts could be replaced by an equivalent Turing machine program.

The effect of each machine is given by an expression of the form $\alpha \vdash^* \beta$, where α gives the general form of the input tape and β gives the general form of the output tape, that is, the tape resulting after the machine has halted. Since we are using flow charts the instructions are not labeled, so that we indicate the scanned square in α and β by a short underline _. Variable symbols x, y, x_1, y_1, and so forth, always represent words from A^*. If a variable symbol x is underlined (that is, $\underline{x}$), it is to be understood that the leftmost symbol of x is being scanned. If $x = \lambda$, then the symbol immediately to the right of x is being scanned. This is usually a blank.

Since we will be interested in standard computability, it is important to ensure that the portion of the tape to the left of the initially scanned symbol is unchanged. This portion is represented by $\varphi *$, where φ is a word over $A \cup \{*\}$. We denote the symbols of A by 1, 2, $\ldots$, m (see Table IV.1).

Exercises

1. Construct a Turing machine program which is equivalent to the Plus1 flow chart, for $m = 2$.
2. Construct flow charts for the Turing machines constructed in the exercises of Chapter II.

TABLE IV.1   BASIC MACHINES

Name	Flow Chart	Effect
$R(b)$:	$\overset{\neq b}{\overset{\frown}{R}}\ H$	Find the first occurrence of the symbol b to the right; that is, $$\cdots a\,x\,b\cdots \vdash^* \cdots a\,x\,\underline{b}\cdots$$ (if b is not in x)
$L(b)$:	$\overset{\neq b}{\overset{\frown}{L}}\ H$	Find the first occurrence of the symbol b to the left; that is, $$\cdots b\,x\,a\cdots \vdash^* \cdots \underline{b}\,x\,a\cdots$$ (if b is not in x)
$\text{Move}_n(a)$:	$*\ R(*)^{n+1}a\ L(*)^{n+1}a$	Write "a" $n+1$ blocks to the right; that is, $$\varphi * \underline{a}\,x_1 * x_2 * \cdots * x_n * y * \vdash^*$$ $$\varphi * \underline{a}\,x_1 * x_2 * \cdots * x_n * y\,a$$
COPY_n:		Copy a string n blocks to the right: $$\varphi * \underline{\,}x_1 * x_2 * \cdots * x_n * \vdash$$ $$\varphi * x_1 * \underline{x_2} * \cdots * x_n * x_1$$
CLEAR:		Erase left, to the first blank: $$\varphi * x\,\underline{y}\cdots \vdash^* \varphi * \underline{\ } * {*}^k y\cdots,$$ where $k=\lvert x\rvert$; $y \in A^*$
PLUS1:		Add 1 to y: $$\varphi * y\underline{\,}* \vdash^* \varphi * z\underline{\,}*,$$ where $z = y+1$ in $m$adic notation

TABLE IV.1—*Continued*

Name	Flow Chart	Effect

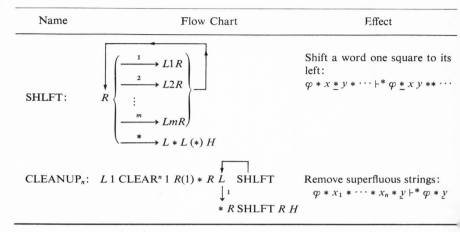

SHLFT: (Flow Chart) Shift a word one square to its left:
$$\varphi * x * \underline{y} * \cdots \vdash^* \varphi * x \underline{y} * * \cdots$$

CLEANUP$_n$: $L\,1\,\text{CLEAR}^n\,1\,R(1) * R\,L$ SHLFT   Remove superfluous strings:
$\downarrow 1$
$* R\,\text{SHLFT}\,R\,H$

$$\varphi * x_1 * \cdots * x_n * \underline{y} \vdash^* \varphi * \underline{y}$$

2. STANDARD COMPUTABILITY

The following definitions will play a central role in these notes.

DEFINITION IV.2.1

A partial function $f : (A^*)^n \to A^*$ is **standard computable** if and only if there is a Turing machine $Z = (A, P)$ with initial label p and final label q such that

(a) if $f(\vec{x}_n) = y$ and if $w * px_1 * x_2 * \cdots * x_n$ is an I.D. for some word w over $A \cup \{*\}$, then $w * px_1 * \cdots * x_n \vdash^*_Z w * qy$; and
(b) if $f(\vec{x}_n)$ is undefined, then Z loops when given any I.D. of the form $w * px_1 * \cdots * x_n$.

If $R(\vec{x}_n)$ is any predicate whose variables range over A^*, we define its **partial characteristic function** $C^p_R(\vec{x}_n)$ and its **total characteristic function** $C^t_R(\vec{x}_n)$ by

$$C^p_R(\vec{x}_n) = \begin{cases} \lambda & \text{if} \quad R(\vec{x}_n) \text{ is true,} \\ \text{undefined} & \text{if} \quad R(\vec{x}_n) \text{ is false;} \end{cases}$$

and

$$C^t_R(\vec{x}_n) = \begin{cases} \lambda & \text{if} \quad R(\vec{x}_n) \text{ is true,} \\ 1 & \text{if} \quad R(\vec{x}_n) \text{ is false.} \end{cases}$$

We say that $R(\vec{x}_n)$ is **standard acceptable** if $C_R^p(\vec{x}_n)$ is standard computable, and **standard decidable** if $C_R^t(\vec{x}_n)$ is standard computable. ∎

Less formally, suppose that Z is as above and that it is started on a tape which contains $w * x_1 * \cdots * x_n$, and that Z is scanning the leftmost symbol of $x_1 * \cdots * x_n$. In other words the tape contains the initial configuration, plus extraneous material to its left. In this case Z must proceed to compute $f(\vec{x}_n)$ [or accept or decide $R(\vec{x}_n)$] as usual, except that it may not disturb the portion of the tape which contains $w *$. Further, when Z stops (if it stops), it must be scanning the symbol immediately to the right of $w *$. If Z decides R and $R(\vec{x}_n)$ is false, the symbol scanned must be 1.

Note that these definitions are compatible with those of Section II.2. Their purpose is to simplify the constructions involving Turing machines which appear in the next section.

3. TURING MACHINE CLOSURE PROPERTIES

We now show that the standard Turing-decidable predicates are closed under the S-rudimentary operations, and establish several other closure properties.

LEMMA IV.3.1

The predicate $x = y$ is standard decidable.

PROOF

For each $a \in A$ define CHECK(a) by the flow chart:

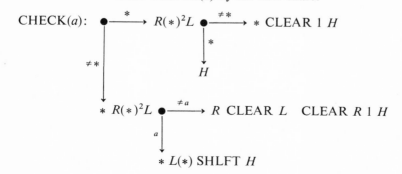

The description in Table IV.2 of the behavior of CHECK is easily verified; a is any symbol of A, b denotes any symbol of A other than a,
and x, y denote words over A.

TABLE IV.2

Tape at Start		Tape after CHECK(a) Halts	
$\varphi * x a * y a *$	$\vdash^*$	$\varphi * \underline{x} * y$	
$\varphi * x a * y b *$	$\vdash^*$	$\varphi * \underline{1}$	
$\varphi * x a **$	$\vdash^*$	$\varphi * \underline{1}$	
$\varphi * \underline{*} * y *$	$\vdash^*$	$\varphi * \underline{1}$	if $y \neq \lambda$
$\varphi \underline{*} **$	$\vdash^*$	$\varphi * \underline{*}$	

The following machine "*Eq*" standard decides the predicate $x = y$:

$$Eq : L\ R(*)\ L \left\{ \begin{array}{l} \xrightarrow{\ 1\ } \text{CHECK(1)} \\ \xrightarrow{\ 2\ } \text{CHECK(2)} \\ \quad\vdots \\ \xrightarrow{\ m\ } \text{CHECK(}m\text{)} \\ \xrightarrow{\ *\ } \text{CHECK(1)}\ H \end{array} \right\} \xrightarrow{*}\ \xrightarrow{\ 1\ } H$$

COROLLARY IV.3.2

The catenation predicate $xy = z$ is standard decidable.

PROOF

Consider the following flow chart:

$$CAT : L\ R(*)\ \text{SHLFT}\ R(*)\ R\ \text{SHLFT}\ L(*)\ R\ Eq\ H$$

This machine simply transforms $\varphi * \underline{x} * y * z$ into $\varphi * \underline{xy} * z$,
and then applies Eq. ∎

LEMMA IV.3.3

If $R(\vec{x}_n)$ is standard decidable, then $\neg\ R(\vec{x}_n)$ is also standard decidable.

PROOF

Let $Z = (A, P)$ be a Turing machine which standard decides R. Then the machine corresponding to the following flow chart standard decides $\neg R$:

$$P \bullet \xrightarrow{\ 1\ } * H$$
$$\Big\downarrow *$$
$$1\ H.$$

This machine merely replaces 1 by $*$ and $*$ by 1. ∎

LEMMA IV.3.4

If $Q(\vec{x}_n)$ and $R(\vec{x}_n)$ are standard decidable, then so are $Q(\vec{x}_n) \wedge R(\vec{x}_n)$ and $Q(\vec{x}_n) \vee R(\vec{x}_n)$.

PROOF

Let $Z_q = (A, P_q)$ and $Z_r = (A, P_r)$ be Turing machines which standard decide Q and R, respectively. Then the following flow chart describes a Turing machine $Z = (A, P)$ which standard decides $Q \wedge R$.

$$P: L\ COPY_n^n\ R\ P_q\ \bullet \xrightarrow{\ *\ } L\ L(*)^n\ R\ P_r\ H$$
$$\Big\downarrow 1$$
$$CLEANUP_n\ H.$$

For a given input $x_1, \ldots, x_n \in A^*$, this machine first copies $x_1, \ldots, x_n$, producing $x_1 * \cdots * x_n * x_1 * \cdots * x_n$. Machine Z_q is applied to the second copy; since it standard decides Q, it does not disturb the first $x_1 * \cdots * x_n$. Thus either the tape $x_1 * \cdots * x_n * *$ or the tape $x_1 * \cdots * x_n * 1$ results, depending on whether $Q(\vec{x}_n)$ is true or false. If true, the left end of the first copy is found and Z_r is applied to $x_1 * \cdots * x_n$. Machine Z then accepts the input or not, according as $R(\vec{x}_n)$ is true or false. If $Q(\vec{x}_n)$ is false, then Z rejects its input immediately.

Finally, $Q \vee R \Leftrightarrow \neg(\neg Q \wedge \neg R)$, so that closure under $\neg$ and $\wedge$ (already proved) implies closure under $\vee$. ∎

LEMMA IV.3.5

If $R(\vec{y}_m)$ is standard decidable and $Q(\vec{x}_n)$ is an explicit transform of R, then Q is standard decidable.

PROOF

$Q(\vec{x}_n) \Leftrightarrow R(\xi_1, \ldots, \xi_m)$, where each ξ_i is either in A^* or is one of $x_1, \ldots, x_n$. Suppose that R is standard decidable by Turing machine $Z_r = (A, P_r)$. We construct a machine $Z_q = (A, P_q)$ which will first transform a tape $\varphi \underline{*} x_1 * \cdots * x_n$ into the tape $\varphi \underline{*} x_1 * \cdots * x_n * \xi_1 * \cdots \xi_m$, and then will apply Z_r to $\xi_1, \ldots, \xi_m$. This will give $\varphi * x_1 * \cdots * x_n * \underline{1}$ or $\varphi * x_1 * \cdots * x_n *\underline{*}$, which is immediately changed to $\varphi * \underline{1}$ or $\varphi *\underline{*}$, respectively. Z_q is given as follows:

First, define machine $T = T_1 T_2 \cdots T_m$, where each T_i will convert the tape $\varphi \underline{*} x_1 * \cdots * x_n * \xi_1 * \cdots * \xi_{i-1}$ into $\varphi \underline{*} x_1 * \cdots * x_n * \xi_1 * \cdots * \xi_i$. For each $i = 1, 2, \ldots, m$, define T_i by the following:

(a)   If ξ_i is the variable x_j, then T_i is given by

$$T_i : R(*)^{j-1} \, \mathrm{COPY}_{n+i-j} \, L(*)^j.$$

(b)   If $\xi_i = b_1 \cdots b_k$ and $b_1, \ldots, b_k \in A$, then T_i is given by

$$T_i : R(*)^{n+i} b_1 R \, b_2 R \cdots R b_k \, L(*)^{n+i}.$$

Clearly the T_i behave as required, and T converts $\varphi \underline{*} x_1 * \cdots * x_n$ into $\varphi \underline{*} x_1 * \cdots * x_n * \xi_1 * \cdots * \xi_m$. Finally, Z_q is given by

$$Z_q : L \, T \, R(*)^n \, R \, Z_r, \, \mathrm{CLEANUP}_n \, H. \qquad \blacksquare$$

LEMMA IV.3.6

If $R(\vec{x}_n, y)$ is standard decidable, then $\exists y R(\vec{x}_n, y)$ is standard acceptable.

PROOF

Let R be standard decided by Turing machine $Z_r = (A, P_r)$. We construct machine $Z = (A, P)$, where P is given by the following flow chart:

$$P : L \, \mathrm{COPY}_{n+1}^{n+1} \, R \, P_r \xrightarrow{\ 1\ } * \, L \, \mathrm{PLUS1} \, L(*)^{n+1}$$

$$\downarrow *$$

$$\mathrm{CLEANUP}_{n+1} \, H.$$

The input to P is $\varphi * \underline{x}_1 * \cdots * x_n *$, which is of course identical to $\varphi * \underline{x}_1 * x_2 * \cdots * x_n **$, the tape corresponding to $(x_1, \ldots, x_n, \lambda)$. Now suppose that for some y, that the tape contains $\varphi \underline{*} x_1 * \cdots * x_n * y$. P first copies this tape, yielding $\varphi * x_1 * \cdots * x_n * y \underline{*} x_1 * \cdots * x_n * y$. P_r is then applied to the rightmost copy, yielding $\varphi * x_1 * \cdots * x_n * y *\underline{*}$ if $R(\vec{x}_n)$ is true, and $\varphi * x_1 * \cdots * x_n * y * \underline{1}$ if false. If false, P erases the $\underline{1}$ and replaces y by its successor y' in the $m$adic ordering of A^*, yielding $\varphi * \underline{x}_1 * \cdots * x_n * y'$. This cycle repeats as long as $R(\vec{x}_n, y)$ is false, as y ranges over A^*. If at some point $R(\vec{x}_n, y)$ is found to be true, P clears the tape and stops in standard position, thus accepting its input.

Clearly P will stop if and only if there is some y for which $R(\vec{x}_n, y)$ is true. ∎

COROLLARY IV.3.7

If $R(\vec{x}_n, y)$ is standard decidable, then $\mu y R(\vec{x}_n, y)$ is standard computable.

PROOF

We merely modify machine P above to stop with y on its tape when the first y is found for which $R(\vec{x}_n, y)$ is true, as follows:

$$P' : L \, \mathrm{COPY}^{n+1}_{n+1} \, R \, P_r \xrightarrow{\quad 1 \quad} * L \, \mathrm{PLUS1} \, L(*)^{n+1}$$

$$\downarrow *$$

$$L \, L(*) \, \mathrm{CLEANUP}_n \, H.$$

As before, P' eventually yields $\varphi * x_1 * \cdots * x_n * \underline{y} * *$, where y is the least y such that $R(\vec{x}_n, y)$ is true. P' then uses CLEANUP to obtain $\varphi * \underline{y}$. ∎

NOTATION

We shall use the following abbreviations, where $\leq$ is interpreted numerically:

$$(\exists z)_{\leq y} R(\vec{x}_n, z) \qquad \text{for} \quad \exists z (z \leq y \wedge R(\vec{x}_n, z));$$

and

$$(\forall z)_{\leq y} R(\vec{x}_n, z) \qquad \text{for} \quad \forall z (z \leq y \Rightarrow R(\vec{x}_n, z)).$$

LEMMA IV.3.8

If $R(x_n, y)$ is standard decidable, then so are $(\exists z)_{\leq y} R(\vec{x}_n, z)$ and $(\forall z)_{\leq y} R(\vec{x}_n, z)$.

PROOF

Define $f(\vec{x}_n, y)$ by

$$f(\vec{x}_n, y) = \mu z\big(R(\vec{x}_n, y) \vee z = y\big).$$

Now, the predicate $R(\vec{x}_n, z) \vee z = y$ [call it $Q(\vec{x}_n, y, z)$] is constructed from R and $z = y$ by disjunction and explicit transformations and we know by Lemma IV.3.1 that $z = y$ is standard decidable. Thus, by Lemmas IV.3.4 and IV.3.5 Q is standard decidable, so that f is standard computable by the previous lemma. Suppose $Z_f = (A, P_f)$ standard computes f.

We can decide whether or not $\exists z(z \leq y \wedge R(\vec{x}_n, z))$ is true simply by checking to see whether $R(\vec{x}_n, f(\vec{x}_n, y))$ is true (verify this statement). The following machine standard decides $(\exists z)_{\leq y} R(\vec{x}_n, z)$ in this way. Let P_r be a machine which standard decides R.

$$Z : L \text{ COPY}_{n+1}^{n+1} R P_f \text{ CLEANUP}_1 L(*)^{n+1} R P_r H.$$

Z first converts the input tape $\varphi * \underline{x_1} * \cdots * x_n * y$ to $\varphi * x_1 * \cdots * x_n * y * \underline{x_1} * \cdots * x_n * y$, and then applies P_f, yielding $\varphi * x_1 * \cdots * x_n * y * \underline{f(\vec{x}_n, y)}$. Z converts this to $\varphi * \underline{x_1} * \cdots * x_n * f(\vec{x}_n, y)$, and then applies P_r, which standard decides $R(\vec{x}_n, f(\vec{x}_n, y))$.

Thus $(\exists z)_{\leq y} R(\vec{x}_n, z)$ is standard decidable. This immediately implies that $(\forall z)_{\leq y} R(x_n, z)$ is standard decidable by Lemma IV.3.3, since it is equivalent to $\neg (\exists z)_{\leq y} \neg R(\vec{x}_n, y)$. ∎

These operations are called **bounded existential** and **bounded universal quantification**, respectively.

COROLLARY IV.3.9

If $R(\vec{x}_n, y)$ is standard decidable, then so are $(\exists z)_{\mathbf{P}y} R(\vec{x}_n, z)$ and $(\forall z)_{\mathbf{P}y} R(\vec{x}_n, z)$.

PROOF

First note the following:

$$x\mathbf{P}y \Leftrightarrow (\exists u)_{\leq y}(\exists v)_{\leq y}(\exists w)_{\leq y}[v = ux \wedge y = vw].$$

By the previous lemma $x\mathbf{P}y$ is standard decidable. Now note that

$$(\exists z)_{\mathbf{P}y} R(\vec{x}_n, z) \Leftrightarrow (\exists z)_{\leq y}(z\mathbf{P}y \wedge R(\vec{x}_n, z))$$

and

$$(\forall z)_{\mathbf{P}y} R(\vec{x}_n, z) \Leftrightarrow (\forall z)_{\leq y}(z\mathbf{P}y \Rightarrow R(\vec{x}_n, Z)).$$

Again by the previous lemma, the right-hand sides are standard decidable.   ∎

THEOREM IV.3.10

Every S-rudimentary predicate is standard decidable.

PROOF

By Lemma IV.3.2 the basis predicate $xy = z$ is standard decidable. By Lemmas IV.3.3, 3.4, 3.5, and 3.9, the class of standard decidable predicates is closed under the S-rudimentary operations.   ∎

THEOREM IV.3.11

Every existentially definable predicate is standard acceptable.

PROOF

Immediate from Lemma IV.3.6 and the previous theorem.   ∎

LEMMA IV.3.12

Every S-rudimentary function is standard computable.

PROOF

Let $h(\vec{x}_n)$ be S-rudimentary. Then for all $x \in A^*$,

$$h(\vec{x}_n) = \mu y(h(\vec{x}_n) = y).$$

By definition, $h(\vec{x}_n) = y$ is S-rudimentary, and so is standard decidable by Theorem IV.3.10. The result now follows from Corollary IV.3.7.   ∎

LEMMA IV.3.13

If $h(\vec{x}_p)$ is the composition of $f(\vec{x}_n)$ with $g_1(\vec{y}_p)$, ..., $g_n(\vec{y}_p)$ and $f, g_1, g_2, \ldots, g_n$ are all standard computable functions, then h is also standard computable.

PROOF

Let f, g_1, g_2, ..., g_n be standard computed by Turing machines Z_f, Z_1, ..., Z_n with flow charts P_f, P_1, P_2, ..., P_n, respectively. For each value of $i = 1, 2, \ldots, n$, define machine T_i by the following diagram:

$$T_i : \text{COPY}_{p+i-1}^p \, RR(*)^{i-1} R \, P_i \, L(*)^{p+i}.$$

It is easily verified that T_i will convert a tape of the form $\varphi * y_1 * \cdots * y_p * z_1 * \cdots * z_{i-1} *$ to $\varphi * y_1 * \cdots * y_p * z_1 * \cdots * z_i *$, where $z_i = g_i(\vec{y}_p)$. This implies that the sequence $T_1 \, T_2 \, \cdots \, T_n$ will convert $\varphi * y_1 * \cdots * y_p$ into $\varphi * y_1 * \cdots * y_p * z_1 * \cdots * z_n$, where $z_1 = g_1(\vec{y}_p)$, $z_2 = g_2(\vec{y}_p)$, ..., $z_n = g_n(\vec{y}_p)$. Finally we exhibit the following machine, which we claim standard computes $h(\vec{y}_p)$

$$L \, T_1 \, T_2 \, \cdots \, T_n \, R(*)^p \, R \, P_f \, \text{CLEANUP}_p \, H.$$

The instructions through "P_f" convert the input tape $\varphi * y_1 * \cdots * y_p$ into $\varphi * y_1 * \cdots * y_p * \underline{h}(\vec{y}_p)$, which is then changed to $\varphi * \underline{h}(\vec{y}_p)$ by "CLEANUP."   ∎

THEOREM IV.3.14

Every μ-rudimentary function is standard computable.

PROOF

Let f be μ-rudimentary, and let $h(y)$ and $Q(\vec{x}_n, y)$ be as in Definition III.5.1. By Lemma IV.3.12, $h(y)$ is a standard computable function. Further, $\mu y Q(\vec{x}_n, y)$ is standard computable by Corollary IV.3.7. Hence,

$$f(\vec{x}_n) = h\big(\mu y Q(\vec{x}_n, y)\big)$$

is the composition of two standard computable functions and hence is standard computable by the previous lemma.   ∎

Exercises

1. Show that $P(\vec{x}_n) \Rightarrow Q(\vec{x}_n)$ is standard decidable if P and Q are.
2. Show that if $R(\vec{x}_n, y)$ is a standard decidable predicate and $f(\vec{x}_n)$ is a standard computable total function, then the following predicate is also standard decidable:

$$(\exists z)_{\leq f(\vec{x}_n)} \, R(\vec{x}_n, z).$$

4. AUXILIARY TAPE SYMBOLS

It would seem to be quite convenient to provide a Turing machine with extra tape symbols in addition to those of its input alphabet, so that it could use these symbols as special markers during its computation.

In this section we show that such an additional capability does not in fact increase the computational power of Turing machines. This result will be used extensively in Chapter VI; however, it is not requisite to Chapter V, so the reader may wish to skip ahead to Chapter V, and return to this section later.

An Application

As an (informal) example of the utility of extra tape symbols, we can show that with the use of a single auxiliary symbol Δ it is possible to replace an arbitrary Turing machine Z by another machine Z'' which leaves its result in standard form (without imbedded blanks) and which does not disturb the tape contents to the left of the input n-tuple $x_1 * \cdots * x_n$. We first construct a Turing machine Z' which will take a tape of the form $\varphi\Delta\alpha\Delta$ into $\varphi\Delta\beta\Delta$ iff $\alpha \vdash_Z^* \beta$. Thus Z' simulates Z exactly, but operates within end markers, and leaves undisturbed the tape contents to the left of the end markers.

Z' is obtained by replacing each "R" or "L" instruction of Z by an appropriate sequence which will check to see if Z is scanning the end of its tape (by seeing if "Δ" is on an adjacent square). If not, Z' merely copies the action of Z. If Z is about to move beyond the end of α on its tape, Z' will increase the length of α by moving $\alpha\Delta$, or the rightmost Δ, right one position, leaving φ untouched. All this can clearly be done with our library of basic Turing machines.

In this way we can construct a machine Z'' which standard computes $\psi_Z^n(\vec{x}_n)$ (that is, the function which Z computes). Z'' will use one auxiliary symbol Δ, and proceeds as follows:

1. Given an input $\varphi * \underline{x}_1 * \cdots * x_n$, Z'' converts this to
   $\varphi\Delta_1 \underline{x}* \cdots * x_n \Delta$

2. Z'' then applies Z', yielding a tape of the form $\varphi \Delta z \Delta$, whose content is $\psi_Z^n(\vec{x}_n)$ (call this y).

3. Z'' then shifts y to the left one symbol at a time until the Δ is encountered. This gives $\varphi \Delta y w \Delta$ for some w which either starts with $*$, or is empty.

4. Z'' then erases w and the two Δs, and stops on the leftmost symbol of y.

The equivalence of standard computability and ordinary computability will also come as a corollary to one of the main theorems of Chapter V. However, that argument is rather complex and certainly does not provide a practical algorithm to go from a Turing machine which computes $f(x)$ to one which standard computes $f(x)$.

An Encoding Technique

We shall encode the symbols of a large alphabet $A = \{a_1, \ldots, a_m\}$ as words over the alphabet $\{a_1, *\}$. This code h will be extended to words, so if w is a word over $A \cup \{*\}$, $h(w)$ will be a word over $\{a_1, *\}$. If a Turing machine Z is given which uses auxiliary symbols, its computation can be simulated by Turing machine Z as follows:

1. Given input tape $x_1 * \cdots * x_n$, Z' will encode this, obtaining $h(x_1 * \cdots * x_n)$.

2. If $upv \vdash_Z xqv$ is a single step in a computation by Z, then $h(u)ph(v) \vdash_{Z'}^* h(x)qh(y)$ will be true by a series of steps by Z'.

3. Suppose Z halts with I.D. $upv*w(v \in A^*$ and $u, w \ (A \cup \{*\})^*)$, so its content is v. Then Z' will obtain I.D. $h(u)ph(v)h(* w)$, and then decode v, obtaining a final I.D. $h(u)p'v * z \ (z \in (A \cup \{*\})^*)$ whose content is also v.

DEFINITION IV.4.1

Let $A = \{a_1, \ldots, a_m\}$ be an alphabet, " $*$ ". The definition of the code $h:(A \cup \{*\})^* \to \{*, a_1\}^*$ is as follows:

(1) $h(a_i) = *^{i-1}a_1*^{m-i}$ for $i = 1, \ldots, m$,

(2) $h(*) = *^m$,

and

$$(3) \quad h(xa) = h(x)h(a) \qquad \text{for} \quad a \in A \cup \{*\}, \, x \in (A \cup \{*\})^*.$$

Let $\alpha = xpy$ be an I.D. of a Turing machine $Z = (A, P)$, where $x, y \in (A \cup \{*\})^*$. The h-**encoding** of α is defined to be $h(\alpha) = h(x)ph(y)$.

LEMMA IV.4.2

Let $n \geq 1$, $A = \{a_1, \ldots, a_m\}$ and $B = \{a_1, \ldots, a_p\}$, where $p \leq m$ (so $B \subseteq A$). There is a Turing machine $Z_n^0 = (B, P^0)$ such that for any $x_1, \ldots, x_n \in B^*$

$$\underline{x}_1 * \cdots * x_n \vdash^* \underline{h}(x_1 * \cdots * x_n).$$

PROOF

We prove the case $n = 1$; the general case is quite similar, but notationally more complex. For each $i = 1, 2, \ldots, p$, define the machine "CODE$_i$" by:

$$\text{CODE}_i : * L(*) \, \text{SHLFT}^{m-1} R(*)^{i+1} a_1 \, L(*)^i \, H.$$

It is easily seen that the effect of CODE$_i$ is to transform

$$x\underline{a}_i * y \quad \text{into} \quad x \underline{*} h(a_i)y \quad \text{for} \quad a_i \in B, \, x \in B^*, \, y \in A^*.$$

Now Z_n^0 can be defined by

$$Z_n^0 : \quad L \, R(*) \, L \left\{ \begin{array}{l} \xrightarrow{\,a_1\,} \text{CODE}_1 \\ \vdots \\ \xrightarrow{\,a_p\,} \text{CODE}_p \\ \xrightarrow{\,*\,} R \, H. \end{array} \right.$$

Let $x = a_{i_1} \cdots a_{i_k}$. Then Z_n^0 performs the following computation:

$$\underline{a}_{i_1} \cdots a_{i_k} \vdash * a_{i_1} \cdots a_{i_k} \underline{*} \vdash * a_{i_1} \cdots a_{i_{k-1}} \underline{*} h(a_{i_k})$$
$$\vdash * \cdots \vdash * a_{i_1} \underline{*} h(a_{i_2} \cdots a_{i_k})$$
$$\vdash^* \underline{*} h(a_{i_2} \cdots a_{i_k}) \vdash \underline{h}(a_{i_1} \cdots a_{i_k}). \qquad \blacksquare$$

Observe that Z_n^0 uses the symbols of B, and no symbols of $A - B$. This is necessary to encode a large alphabet into $\{a_1, *\}$.

LEMMA IV.4.3

Let A, B, and h be as above, and let $Z = (A, P)$ be a Turing machine. Then there is a Turing machine $Z^1 = (B, P^1)$ such that if α and β are I.D.s of Z, then $\alpha \vdash_Z^* \beta$ if and only if $h(\alpha) \vdash_{Z^1}^* h(\beta)$.

PROOF

P^1 is obtained from P by replacing each instruction in P by a sequence in P^1, as in Table IV.3.

TABLE IV.3

Instruction of P	Sequence in P^1
a_i (write a_i)	$(*R)^{i-1} a_1 (*R)^{m-i} L^m$
R	R^m
L	L^m
$Ja_i p_j$	$R^i \bullet \xrightarrow{\ a_1\ } L^i Jp_j$
	$\downarrow$
	$L^i H$

It is easily verified that Z^1 behaves as required.   ∎

LEMMA IV.4.4

Let A, B, and h be as above. There is a Turing machine $Z^2 = (B, P^2)$ such that if Z^2 is begun with I.D. $h(u)ph(v*w)$, where $v \in B^*$ and u, w are words over $B \cup \{*\}$, then Z^2 will halt with an I.D. $h(u)pv*y$, such that y is a word over $B \cup \{*\}$ and q is the final label of Z^2.

PROOF

Left as an exercise for the reader. What is required is a program which will find $h(v)$, and then decode it by the h function, obtaining an I.D. of the form $h(u)pv*y$.

Auxiliary Symbols Theorem

THEOREM IV.4.5

Let $B \subseteq A$, and suppose $f : (A^*)^n \to B^*$ is a partial function. Let $Z = (A, P)$ be a Turing machine which computes the function f. Then there is a Turing machine $Z' = (B, P')$ which computes the function g defined by:

$$g(\vec{x}_n) = \begin{cases} f(\vec{x}_n) & \text{if } x_1, \ldots, x_n \in B^*, \\ \text{undefined} & \text{if } x_i \in A^*/B^* \text{ for some } i. \end{cases}$$

PROOF

Z' is obtained by combining Z_n^0, Z^1, and Z^2 from the previous lemmas. Let $x_1, \ldots, x_n \in B^*$. Then Z_n^0 will convert the I.D. $p_0 x_1 * \cdots * x_n$ into $p_1 h(x_1 * \cdots * x_n)$. Z^1 will now simulate Z, and if $f(\vec{x}_n)$ is defined Z^1 will halt, obtaining a final I.D. $h(u)ph(v * w)$ such that $v \in B^*$ and $upv * w$ is a final I.D. for Z. Clearly v will equal $f(\vec{x}_n) = g(\vec{x}_n)$, since Z computes g. Finally, Z^2 will decode this, halting with an I.D. whose content is v.

At each step in this process the only symbols used were a_1, $*$, and those of B. Since $a_1 \in B$, Z' is a Turing machine using only alphabet B. ∎

V THE NORMAL FORM
THEOREMS AND
CONSEQUENCES

In this chapter we state and prove the main theorems of the elementary theory of computability. Many of these theorems are straightforward consequences of the normal form theorems, discussed in Section V.1. In a sense the normal form theorems summarize the results of Chapters III and IV. In Chapter III we showed that S-rudimentary and existentially definable predicates can accurately describe Turing machines and their actions; and Chapter IV contained proofs in the other direction, that Turing machines can decide S-rudimentary predicates and accept the existentially definable predicates. As a by-product we obtain a universal Turing machine.

In addition, the machinery which was set up in those chapters is used in Sections V.2 and V.3, to establish many important facts about recursive functions, recursively enumerable sets, and so forth. Many of the arguments used here reflect, in a more formal context, the intuitive arguments which were discussed in Chapter II, Section 1.

106

Section 2 of this chapter contains the main relations which exist among recursive sets, recursively enumerable sets, and recursive functions, and their closure properties. Section V.3 concerns itself with functions, sets and predicates which are uncomputable or undecidable. The main technique used is diagonalization. Section V.4 contains undecidability proofs for more predicates which involve Turing machines, using as a tool the S_n^m theorem of Kleene [K1].

1. THE NORMAL FORM THEOREMS AND A UNIVERSAL TURING MACHINE

THEOREM V.1.1 **(The Normal Form Theorem for Predicates)**

Let A be any alphabet containing at least two symbols, and let $n \geq 1$ be arbitrary. There is an S-rudimentary predicate $T_n(z, \vec{x}_n, y)$ such that, if $P(\vec{x}_n)$ is any $n$ary predicate over A, P is recursively enumerable if and only if there is a word $z_0 \in A^*$ such that

$$P(\vec{x}_n) \Leftrightarrow \exists y\, T_n(z_0, \vec{x}_n, y) \tag{1}$$

is true for all $x_1, \ldots, x_n \in A^*$.

PROOF

Theorem III.4.5 established that T_n is S-rudimentary. The "only if" part of this theorem is immediate from the definition of T_n and is in fact Theorem III.3.2. For the "if," suppose that Equation (1) is valid. Then P is the existential quantification of an S-rudimentary predicate and so is existentially definable. But by Theorem IV.3.11 this implies that P is standard acceptable, so that P is recursively enumerable. ∎

THEOREM V.1.2 **(The Normal Form Theorem for Functions)**

Let A be any alphabet containing at least two symbols, and let $n \geq 1$ be arbitrary. There is an S-rudimentary predicate $T_n(z, \vec{x}_n, y)$ and an S-rudimentary function $U(y)$ such that, if $f(\vec{x}_n)$ is any $n$ary function over A, then f is partial recursive iff there is a word $z_0 \in A^*$ such that for all $x_1, \ldots, x_n \in A^*$ either

$$f(\vec{x}_n) = U\big(\mu y\, T_n(z_0, \vec{x}_n, y)\big) \tag{2}$$

or both sides are undefined.

PROOF

Again "only if" is Theorem III.3.2. Suppose that Equation (2) is true. Then f is by definition μ-rudimentary, since we have shown that U and T_n are S-rudimentary. By Theorem IV.3.14 this implies that f is standard computable and thus partial recursive. ∎

COROLLARY V.1.3

If $\sharp A \geq 2$, a predicate over A is recursively enumerable if and only if it is existentially definable.

COROLLARY V.1.4

If $\sharp A \geq 2$, a function over A is recursive if and only if it is μ-rudimentary.

A Universal Turing Machine

THEOREM V.1.5

For each A such that $\sharp A \geq 2$ and each $n \geq 1$ there is a universal Turing machine $\mathbf{U}_n$ such that for any Turing machine Z there is a word $z_0 \in A^*$ such that for all $x_1, \ldots, x_n \in A^*$:

(i)   If Z eventually halts with final tape content y when given $(x_1, \ldots, x_n)$, then $\mathbf{U}_n$ halts with final tape content y when given $(z_0, x_1, \ldots, x_n)$.

(ii) If Z loops when given $(x_1, \ldots, x_n)$, then $\mathbf{U}_n$ loops when given $(z_0, x_1, \ldots, x_n)$.

PROOF

Define $G(z, \vec{x}_n) = U(\mu y\, T_n(z, \vec{x}_n, y))$. Then $G(z, \vec{x}_n)$ is μ-rudimentary and so is partial recursive. Since G is partial recursive there must be a Turing machine which computes it. Call this machine $\mathbf{U}_n$.

Now let Z be any Turing machine, and let $f(\vec{x}_n)$ be the $n$ary partial function computed by Z. If we let $z_0 = gw(Z)$, we see by the Normal Form Theorem that for every $x_1, \ldots, x_n \in A^*$, either

$$f(\vec{x}_n) = G(z_0, \vec{x}_n),$$

or both sides are undefined. Thus U_n halts when given $(z_0, \bar{x}_n)$ just in case Z halts when given $(\bar{x}_n)$, and with the same final tape contents. ∎

Exercises

1. Using Theorem V.1.2, show that for any Turing machine Z which computes a function $f(x)$, there is another Turing machine Z' which standard computes f. Explain how to obtain Z' effectively from Z.
2. Following is a definition of the nondeterministic Turing machine, the way in which it operates, and the way in which it is used to accept a set. Prove that if a set is accepted by a nondeterministic Turing machine, then it is recursively enumerable.

DEFINITION

A **nondeterministic Turing machine** is a pair $Z = (A, P)$, where A is an alphabet and $P = p_1 : i_1 \cdots p_k : i_k \, p_{k+1}$ is a program exactly as in Definition III.2.1, with the following exception: any instruction may be of the form,

$$Jpq, \text{ read as "Jump to label } p \text{ or } q."$$

The terms "instantaneous description" is defined exactly as before. The "immediate consequence" relation ⊢ is defined as in Definition III.2.3, with the following extension:

If $\alpha = up_j av$ is an I.D. and instruction i_j is $Jp_l p_m$, then $\alpha \vdash \beta$ and $\alpha \vdash \gamma$ are *both* true, where

$$\beta = up_l av \qquad \text{and} \qquad \gamma = up_m av.$$

The term "computation," "initial I.D.," and "final I.D." are defined exactly as before. Finally, the **set accepted** by Z is

$\{x \mid$ there is a computation $\alpha_1 \alpha_2, \ldots, \alpha_n$, such that α_1 is the initial I.D. for x, and α_n is final$\}$. ∎

Informally speaking, the nondeterministic machine may "make guesses"; that is, the next I.D. may not be completely determined by the current I.D. Thus the same input word may give rise to many different computations. An input word x is accepted in case it gives rise to at least one terminating computation. The problem is to show that this type of machine is no more powerful than the usual type of Turing machine, at least in terms of accepting sets.

2. POSITIVE CONSEQUENCES

2.1   Recursive and Recursively Enumerable Predicates

We first establish the positive closure properties of each class with respect to the Boolean operations and bounded or unbounded quantification, and then show that the recursive predicates may be characterized in terms of the recursively enumerable predicates.

THEOREM V.2.1

The class of recursive predicates is closed under conjunction, disjunction, negation, explicit transformation, and bounded quantification.

PROOF

Immediate from Lemmas IV.3.3, IV.3.4, IV.3.5, and IV.3.8.   ∎

THEOREM V.2.2

The class of recursively enumerable predicates is closed under conjunction, disjunction, explicit transformation, bounded and subword existential and universal quantification, and existential quantification.

PROOF

By Corollary V.1.3 it suffices to show that the class of existentially definable predicates has these closure properties. Explicit transformation is immediate, so we proceed to conjunction and disjunction. Suppose $R(\vec{x}_n)$ and $S(\vec{x}_n)$ are existentially definable. Then there must be S-rudimentary predicates Q_r and Q_s such that the following hold for all $x_1, \ldots, x_n \in A^*$:

$$R(\vec{x}_n) \Leftrightarrow \exists u Q_r(\vec{x}_n, u) \tag{1}$$

$$S(\vec{x}_n) \Leftrightarrow \exists v Q_s(\vec{x}_n, v). \tag{2}$$

We then have the following

$$R(\vec{x}_n) \vee S(\vec{x}_n) \Leftrightarrow \exists w \big(Q_r(\vec{x}_n, w) \vee Q_s(\vec{x}_n, w) \big) \tag{3}$$

$$R(\vec{x}_n) \wedge S(\vec{x}_n) \Leftrightarrow \exists w (\exists u, v)_{\mathbf{P}w} \big(Q_r(\vec{x}_n, u) \wedge Q_s(\vec{x}_n, v) \big). \tag{4}$$

That Formula (3) holds is immediate. If the right side of Formula (4) holds, then the left side is true. If the left side is true then let $w = uv$, where u and v are as in Formulas (1) and (2). Then the right-hand side of Formula (4) is also true, so the equivalence holds. The formulas following "$\exists w$" are S-rudimentary; thus $R \vee S$ and $R \wedge S$ are existentially definable.

For the quantifiers, suppose that $R(\vec{x}_n, y)$ is existentially definable. There is an S-rudimentary $Q(\vec{x}_n, y, u)$ such that

$$R(\vec{x}_n, y) \Leftrightarrow \exists u Q(\vec{x}_n, y, u). \tag{5}$$

Consider the following formulas:

$$\exists y R(\vec{x}_n, y) \Leftrightarrow \exists w ((\exists y, u)_{\mathbf{P}w} \, Q(\vec{x}_n, y, u)) \tag{6}$$

$$(\exists z)_{\leq y} R(\vec{x}_n, z) \Leftrightarrow \exists w (\exists z)_{\leq y} \, Q(\vec{x}_n, z, w) \tag{7}$$

$$(\forall z)_{\leq y} R(\vec{x}_n, z) \Leftrightarrow \exists w (\forall z)_{\leq y} (\exists u)_{\leq w} \, Q(\vec{x}_n, z, u). \tag{8}$$

To see that Formula (6) holds, suppose that $R(\vec{x}_n, y)$ is true. Let $w = yu$ where u is the word given by Formula (5). Then the right side of Formula (6) is true with this w. The converse is immediate. Formula (7) is very similar. For Formula (8), suppose that $(\forall z)_{\leq y} R(\vec{x}_n, z)$ is true. Let $z_1, z_2, \ldots, z_k$ be an enumeration of all words for which $z \leq y$, and let $u_1, u_2, \ldots, u_k$ be an enumeration of all the corresponding us given by Formula (5). Let $w = u_1 \cdots u_k$. Then the right side of Formula (8) is true for this choice of w. It is immediate that truth of the right side of Formula (8) implies the left. The proof for subword quantification is very similar.

In Formula (6) the portion following "$\exists w$" is S-rudimentary, so $\exists y R$ is existentially definable. In Formulas (7) and (8) Q is S-rudimentary and thus standard decidable (Theorem IV.3.10); by Lemma IV.3.8, the formulas following "$\exists w$" in Formulas (7) and (8) are also standard decidable, and so existentially definable. By the result just shown [involving Formula (6)], the right-hand sides of Formulas (7) and (8) also specify existentially definable predicates. ∎

The proof method of the following is essentially the same as that of Proposition II.1.4.

THEOREM V.2.3

$P(\vec{x}_n)$ is recursive if and only if $P(\vec{x}_n)$ and $\neg P(\vec{x}_n)$ are both recursively enumerable.

PROOF

"Only if" is immediate from Theorem V.2.1. Suppose that P and $\neg P$ are both recursively enumerable. Then by Corollary V.1.3, there must exist S-rudimentary predicates $Q(\vec{x}_n, y)$ and $R(\vec{x}_n, y)$ such that for all $x_1, \ldots, x_n \in A^*$

$$P(\vec{x}_n) \Leftrightarrow \exists y \, Q(\vec{x}_n, y)$$
$$\neg P(\vec{x}_n) \Leftrightarrow \exists y \, R(\vec{x}_n, y).$$

Define the function $f(\vec{x}_n)$ by

$$f(x_n) = \mu y \left(Q(\vec{x}_n, y) \vee R(\vec{x}_n, y) \right).$$

Now f is clearly μ-rudimentary and so is recursive. Since one of $P(\vec{x}_n)$ and $\neg P(\vec{x}_n)$ is true for each $x_1, \ldots, x_n \in A^*$, f is also total. Let $C_P^t(\vec{x}_n)$ and $C_Q^t(\vec{x}_n, y)$ be the total characteristic functions of P and Q, respectively (as in Definition IV.2.1). Now Q is recursive so by definition we know that C_Q^t is a total recursive function. But this immediately implies that C_P^t is also total recursive by Lemma IV.3.13, as follows;

$$C_P^t(\vec{x}_n) = C_Q^t(\vec{x}_n, f(\vec{x}_n))$$
$$\text{for all} \quad x_1, \ldots, x_n \in A^*.$$

Thus P is a recursive predicate. ∎

2.2   Relations to Recursive Functions

THEOREM V.2.4

Let $f(\vec{x}_n)$ be a partial function.

(a) If f is recursive then the predicate $f(\vec{x}_n) = w$ is recursively enumerable.

(b) If the domain of f is recursive and the predicate $f(\vec{x}_n) = w$ is recursively enumerable, then f is a recursive function.

PROOF

(a) By the Normal Form Theorem there is a word $z_0 \in A^*$ such that $f(\vec{x}_n) = U(\mu y \, T_n(z_0, \vec{x}_n, y))$ for all $x_1, \ldots, x_n \in A^*$. This immediately implies that

$$f(\vec{x}_n) = w \Leftrightarrow \exists y \left(U(y) = w \wedge T_n(z_0, \vec{x}_n, y) \right).$$

Now U and T_n are both S-rudimentary; thus $f(\vec{x}_n) = w$ is existentially definable and so is recursively enumerable.

(b) Now let $f(\vec{x}_n) = w$ be recursively enumerable, and let $D(\vec{x}_n)$ be true iff $f(\vec{x}_n)$ is defined. Then $f(\vec{x}_n) \neq w$ is recursively enumerable by Theorem V.2.2 and

$$f(\vec{x}_n) \neq w \Leftrightarrow \neg D(\vec{x}_n) \vee \exists y(f(\vec{x}_n) = y \wedge y \neq w).$$

Thus by Theorem V.2.3 the predicate $f(\vec{x}_n) = w$ is recursive. By Corollary IV.3.7 this immediately implies that $f(\vec{x}_n)$ is a recursive function, since

$$f(\vec{x}_n) = \mu w(f(\vec{x}_n) = w). \qquad \blacksquare$$

Consequently a total function $f(\vec{x}_n)$ is recursive if and only if the predicate $f(\vec{x}_n) = w$ is recursively enumerable.

We now show that the substitution of a recursive function into a recursively enumerable predicate yields a recursively enumerable predicate.

THEOREM V.2.5

Let $R(\vec{x}_n, y)$ be an $(n + 1)$ary predicate and let $f(\vec{x}_n)$ be an $n$ary function. Then

(a) if $R(\vec{x}_n, y)$ is recursively enumerable and $f(\vec{x}_n)$ is recursive, then $R(\vec{x}_n, f(\vec{x}_n))$ is recursively enumerable; and
(b) if $R(\vec{x}_n, y)$ is recursive and $f(\vec{x}_n)$ is recursive and has a recursive domain, then $R(\vec{x}_n, f(\vec{x}_n))$ is recursive.

PROOF

By Theorems V.2.2 and V.2.4, the following equivalence shows that (a) is true:
$$R(\vec{x}_n, f(\vec{x}_n)) \Leftrightarrow \exists y \left(f(\vec{x}_n) = y \wedge R(\vec{x}_n, y) \right).$$

To prove (b) we need only show that $\neg R(\vec{x}_n, f(\vec{x}_n))$ is also recursively enumerable, by Theorem V.2.3. Let $D(\vec{x}_n)$ be true iff $f(\vec{x}_n)$ is defined. By assumption D is recursive, so that $\neg R(\vec{x}_n, f(\vec{x}_n))$ is recursively enumerable as follows:
$$\neg R(\vec{x}_n, f(\vec{x}_n)) \Leftrightarrow \neg D(\vec{x}_n) \vee \exists y(f(\vec{x}_n) = y \wedge \neg R(\vec{x}_n, y)). \qquad \blacksquare$$

The following theorem resembles Proposition II.1.2 in its intuitive content and proof.

THEOREM V.2.6

A set S is recursively enumerable if and only if S is the range of a recursive function.

PROOF

Let S be the range of a recursive partial function $f(x_1, \ldots, x_n)$. Then we have

$$y \in S \Leftrightarrow \exists \vec{x}_n\, f(\vec{x}_n) = y$$

By Theorem V.2.4 $f(\vec{x}_n) = y$ is recursively enumerable, so that $y \in S$ is recursively enumerable by Theorem V.2.2.

Now suppose that S is recursively enumerable. The case $S = \varnothing$ is trivial, so assume $S \neq \varnothing$. Then there must be a Turing machine $Z = (A, P)$ which accepts S. Let $z_0 = gw(Z)$, and let a be any fixed element of S (there is one, since $S \neq \varnothing$). Consider the formula

$$g(x, y) = w \Leftrightarrow \big(w = x \wedge T_1(z_0, x, y)\big) \vee \big(w = a \wedge \neg T_1(z_0, x, y)\big).$$

The predicate $g(x, y) = w$ is S-rudimentary and so is recursive; further g is total, so by Theorem V.2.4, g is a total recursive function.

It remains to show that S is the range of g. First, if $g(x, y) = w$ then either $w = a$ and so is in S, or $T_1(z_0, x, y)$ is true. In this case Z accepts x, so $w = x$ is again in S.

Conversely, suppose x is an arbitrary element of S. Then Z accepts x, so there is a computation by Z starting with x; thus $T_1(z_0, x, y)$ is true for some y. For this y, $g(x, y) = x$, so x is in the range of g. Thus S equals the range of g.

This result can be strengthened somewhat; that is, S is the range of a *unary* total recursive function f, defined as follows:

$$f(u) = w \Leftrightarrow \big(\neg(\exists\, \sharp)_{Pu}\, \mathrm{Seq}(u, \sharp)\big) \wedge w = a$$
$$\vee\ (\exists\, \sharp, x, y)_{Pu}[\mathrm{Seq}(u, \sharp) \wedge g(x, y) = w$$
$$\wedge\ \mathrm{First}(u, x) \wedge \mathrm{Last}(u, y)]. \qquad\blacksquare$$

THEOREM V.2.7

The following statements about a nonempty set S are equivalent:

   (i) S is recursively enumerable.
  (ii) S is the domain of a partial recursive function.
 (iii) S is the range of a total recursive function.
 (iv) S is the range of a partial recursive function.

PROOF

By definition, (i) and (ii) are equivalent. It is clear that (iii) implies (iv). By the proof just given, (i) implies (iii) and (iv) implies (i). ∎

Similar theorems can be shown for an arbitrary recursively enumerable predicate. In particular, if $P(\vec{x}_n)$ is recursively enumerable, then partial S-rudimentary functions $f_1(u), \ldots, f_n(u)$ can be found such that, for all $x_1, \ldots, x_n \in A^*$,

$$P(\vec{x}_n) \Leftrightarrow \exists u\big(f_1(u) = x_1 \wedge f_2(u) = x_2 \wedge \cdots \wedge f_n(u) = x_n\big).$$

THEOREM V.2.8

For every recursive function $f(\vec{x}_n)$ there is a function $g(\vec{x}_n)$ such that

(i) the predicate $g(\vec{x}_n) = y$ is S-rudimentary, and
(ii) $f(\vec{x}_n) = U\big(g(\vec{x}_n)\big)$ for all $x_1, \ldots, x_n \in A^*$.

PROOF

Let f be computed by Turing machine Z with Gödel word z_0. Define g by: $g(\vec{x}_n) = y \Leftrightarrow T_n(z_0, \vec{x}_n, y)$. Now g is well defined since there is at most one computation beginning with $(\vec{x}_n)$, and the predicate $g(\vec{x}_n) = y$ is S-rudimentary because T_n is S-rudimentary. Condition (ii) above is merely the Normal Form Theorem for functions. ∎

This theorem is interesting because of the simplicity of U: essentially all U does is erase. This can be interpreted as saying that there are functions $g(\vec{x}_n)$ of arbitrary complexity for which $g(\vec{x}_n) = y$ is S-rudimentary, for the following reason. Suppose that $\mathscr{F}$ is any class whatever of recursive functions such that

(i) if $f(\vec{x}_n) \in \mathscr{F}$, then $U\big(f(\vec{x}_n)\big) \in \mathscr{F}$; and
(ii) there is at least one recursive function which is not in $\mathscr{F}$.

Let $f(\vec{x}_n)$ be any recursive function not in $\mathscr{F}$, and let $g(\vec{x}_n)$ be the function given by the previous theorem. Then $g(\vec{x}_n) = y$ is S-rudimentary, but g cannot be in $\mathscr{F}$, for by (i) above this would imply that f was in $\mathscr{F}$, contradicting (ii).

We see from this that it can be much easier to check whether $g(\vec{x}_n) = y$ given $x_1, \ldots, x_n$ and y, than to compute $g(\vec{x}_n)$ given $x_1, \ldots, x_n$ only.

Exercises

1. Show that if $f(\vec{x}_n)$ is a total recursive function and $R(\vec{x}_n, y)$ is a recursive predicate, then the predicate

$$\exists z\bigl(z \leq f(\vec{x}_n) \wedge R(\vec{x}_n, z)\bigr)$$

is also recursive.

2. Show that a set S is recursive if it is the range of a recursive function $f(x)$ such that
    (a) f is monotonic [so $x < y$ implies $f(x) < f(y)$]; or
    (b) $f(x) \geq x$ for all x.

3. Prove or disprove that every recursively enumerable set is the range of
    (a) a partial S-rudimentary function;
    (b) a total S-rudimentary function.

3. NEGATIVE CONSEQUENCES

In this section we prove that a number of functions and predicates are not computable or Turing decidable. Most of the proofs use, either directly or indirectly, the diagonalization methods mentioned in Subsection II.1.3. The following is a formalization of Proposition II.1.12.

THEOREM V.3.1

The predicate $\exists y\, T_1(x, x, y)$ is recursively enumerable but not recursive.

PROOF

T_1 has been shown S-rudimentary, so $T_1(x, x, y)$ and $\exists y\, T_1(x, x, y)$ are both recursively enumerable.

Suppose that $\exists y\, T_1(x, x, y)$ is recursive. By Theorem V.2.3, the unary predicate $\neg \exists y\, T_1(x, x, y)$ is also recursively enumerable, so there is a

Turing machine Z which accepts it. Let $z_0 = gw(Z)$. Then by the Normal Form Theorem for predicates the following is true for all $x \in A^*$:

$$\neg \exists y \, T_1(x, x, y) \Leftrightarrow \exists y \, T_1(z_0, x, y).$$

But this is impossible, since if we let $x = z_0$ the formula becomes

$$\neg \exists y \, T_1(z_0, z_0, y) \Leftrightarrow \exists y \, T_1(z_0, z_0, y).$$

Therefore $\exists y \, T_1(x, x, y)$ cannot be recursive. ∎

COROLLARY V.3.2

The class of all recursively enumerable predicates over A is not closed under negation or the universal quantifier.

PROOF

If the class were closed under negation, $\neg \exists y \, T_1(x, x, y)$ would also be recursively enumerable. But this would imply that $\exists y \, T_1(x, x, y)$ is recursive, which we know to be false.

Now consider $\neg \, T_1(x, x, y)$. This is S-rudimentary since $T_1(z, x, y)$ is S-rudimentary. Again, if the class were closed under universal quantification, then $\neg \exists y \, T_1(x, x, y)$ would be recursively enumerable by the following formula:

$$\neg \exists y \, T_1(x, x, y) \Leftrightarrow \forall y \, \neg \, T_1(x, x, y). \quad ∎$$

The following theorem is closely related to Proposition II.1.8.

THEOREM V.3.3

(a) Suppose $G(k, x)$ is a recursive function which enumerates the set of all unary total recursive functions. Then G is not a recursive function.

(b) The set, S, of Gödel words of Turing machines which compute unary total functions is not recursively enumerable.

PROOF

(a) Suppose G is recursive, and define $h(x) = G(x, x) + 1$. Now G must be total, so h is also total. Further, h is recursive by Lemma IV.3.13; consequently h is in the enumeration given by G, so for some word $u_0 \in A^*$ and all $x \in A^*$

$$h(x) = G(u_0, x).$$

But this is impossible, since this implies

$$G(u_0, u_0) = h(u_0)$$
$$= G(u_0, u_0) + 1 \qquad \text{(by definition of } h\text{)}.$$

(b) Assume that S is recursively enumerable. Then by Theorem V.2.7, S is the range of a total recursive function $g(x)$. Consider the function G defined by

$$G(k, x) = w \Leftrightarrow \exists z, y\, (z = g(k) \wedge w = U(y) \wedge T_1(z, x, y)).$$

It is easily seen that G is total and that w is uniquely determined by k and x; further, G is recursive by Theorem V.2.4.

If $f(x)$ is an arbitrary total function computed by Turing machine Z, then the Gödel word $z_0 = gw(Z)$ must be in S; consequently $z_0 = g(k)$ for some k. Thus for all x

$$G(k, k) = U(\mu y\, T_1(g(k), x, y)) \qquad \text{(by definition of } G\text{)}$$
$$= U(\mu y\, T_1(z_0, x, y))$$
$$= f(x) \qquad \text{(by the Normal Form Theorem)}.$$

Thus G enumerates the set of all total recursive functions. This contradicts (a) above; consequently S cannot be recursively enumerable.   ∎

Suppose $f: (A^*)^n \to A^*$ is a partial function. Define the **completion** of f to be the total function g given by

$$g(\vec{x}_n) = \begin{cases} f(\vec{x}_n) & \text{if } f(\vec{x}_n) \text{ is defined,} \\ \lambda & \text{if } f(\vec{x}_n) \text{ is undefined.} \end{cases}$$

It might seem reasonable to ignore partial functions altogether, and to study only the completions of partial functions and total functions. The following theorem shows that this is impractical, since f may be computable while g is not.

THEOREM V.3.4

There is a partial recursive function whose completion is not recursive.

PROOF

Define $f(x)$ by

$$f(x) = \mu y\, T_1(x, x, y).$$

Then f is partial recursive because it is μ-rudimentary. Further, $f(x) \neq \lambda$ for all $x \in A^*$, because y is the Gödel word of a computation whenever $T_1(x, x, y)$ is true, and such a word must be nonempty. Let $g(x)$ be the completion of $f(x)$, and assume that g is recursive.

Then $g(x) = \lambda$ if and only if there is no y for which $T_1(x, x, y)$ is true; that is,

$$\neg \, \exists y \, T_1(x, x, y) \Leftrightarrow g(x) = \lambda.$$

The right-hand side of this equation is recursively enumerable by Theorem V.2.4. But this is a contradiction, since we have shown that $\neg \, \exists y \, T_1(x, x, y)$ is not recursively enumerable. ∎

Halting Problems

We shall say that a predicate $P(\vec{x}_n)$ is **decidable** if $P(\vec{x}_n)$ is recursive, and **undecidable** if it is not recursive. If P is undecidable, then there is no Turing machine which will always halt for any $(\vec{x}_n)$, over a blank square if $P(\vec{x}_n)$ is true, and over a marked square if $P(\vec{x}_n)$ is false.

If we have an informal predicate P involving Turing machines, instantaneous descriptions, and so forth, we shall say that P is decidable or undecidable just in case the corresponding predicate involving the Gödel words of the Turing machines, and so forth, is decidable or undecidable.

In particular the following informal predicates are of interest:

(a) HALT(Z, x) is true iff Turing machine Z will eventually halt when given input x.

(b) HALT$_Z(x)$ is true iff a fixed Turing machine Z will eventually halt when given input x. Note that HALT$_Z$ is a distinct unary predicate for each Turing machine Z.

(c) PRINT$_a(Z, x)$ is true iff Z will eventually print the symbol a when given input x (however, Z may or may not eventually halt).

(d) EMPTY(Z) is true iff Z accepts no words whatever.

(e) ALL(Z) is true iff Z accepts every word in A^*.

(f) FINITE(Z) is true iff Z accepts only a finite number of words.

(g) EQDMN(Z_1, Z_2) is true iff for every word $x \in A^*$, either Z_1 and Z_2 both halt or they both fail to terminate.

THEOREM V.3.5

Each of the predicates above is undecidable.

PROOF

We actually show that the corresponding predicates involving Gödel words of the Turing machines are not recursive.

(b)   We first show that there is a fixed Turing machine Z_0 for which $\text{HALT}_{Z_0}(x)$ is undecidable. We have shown in Theorem V.3.1 that $\exists y \, T_1(x, x, y)$ is recursively enumerable but not recursive. Let Z_0 be a Turing machine which accepts $\exists y \, T_1(x, x, y)$. Then Z_0 halts for an input $x \in A^*$ iff $\exists y \, T_1(x, x, y)$ is true, so that we have

$$\text{HALT}_{Z_0}(x) \Leftrightarrow \exists y \, T_1(x, x, y) \qquad \text{for all} \quad x \in A^*.$$

If $\text{HALT}_{Z_0}(x)$ were decidable then $\exists y \, T_1(x, x, y)$ would be recursive, which is false by Theorem V.3.1. Thus $\text{HALT}_{Z_0}(x)$ is undecidable.

(a)   For each Z, $\text{HALT}_Z(x)$ is a unary predicate which is an explicit transform of the binary predicate $\text{HALT}(Z, x)$. If $\text{HALT}(Z, x)$ were decidable, then its explicit transform $\text{HALT}_{Z_0}(x)$ would also be decidable, contradicting (a).

The remaining predicates are shown to be undecidable by reduction of the halting problem to each of them. The methods we use require the construction of special-purpose Turing machines, which could be used to decide the halting problem if any of these problems is decidable; and of course this is impossible. For the moment we appeal to the Church–Turing thesis for the effectiveness of the constructions. More formal proof will be given in the next section.

(c)   To show that $\text{PRINT}_a(Z, x)$ is undecidable, let $Z = (A, P)$ be any Turing machine, and let $Z' = (A \cup \{a\}, P')$, where " a " is a symbol not in A, and $P' = P: a \; p_{k+2}$, where $P = p_1 : i_1 \; \cdots \; p_k : i_k \; p_{k+1}$. Clearly Z' will print the symbol a if and only if Z halts, so for this Z' we have

$$\text{PRINT}_a(Z', x) \Leftrightarrow \text{HALT}_Z(x).$$

By (b) we know that there is a Turing machine Z_0 for which $\text{HALT}_{Z_0}(x)$ is undecidable, so $\text{PRINT}_a(Z_0', x)$ is undecidable for this fixed value of a and Z_0'; thus in general $\text{PRINT}_a(Z, x)$ is undecidable.

(d), (e), (f), (g) We show that if any one of these is decidable, then $\text{HALT}(Z, x)$ is also decidable.

Suppose Z and x are given. From Z and x it is easy to construct another machine Z^x which will act as follows for any input $y \in A^*$:

(i)   erase y,
(ii) write x,
(iii) apply Z to x.

Clearly Z^x will accept every word $y \in A^*$ if Z accepts x, and will accept no word if Z does not accept x. The following are true for all $x \in A^*$ and all Turing machines Z:

$$\text{HALT}(Z, x) \Leftrightarrow \neg\, \text{EMPTY}(Z^x)$$
$$\text{HALT}(Z, x) \Leftrightarrow \text{ALL}(Z^x)$$
$$\text{HALT}(Z, x) \Leftrightarrow \neg\, \text{FINITE}(Z^x).$$

If, for example, $\text{ALL}(Z)$ were decidable, the halting problem could be solved as follows:

1. Read Z, x.
2. Construct Z^x.
3. Decide whether $\text{ALL}(Z^x)$ is true; if so answer "Z halts, given x"; otherwise, answer "Z will not halt, given x."

By appealing to the Church–Turing thesis, this implies the existence of a Turing machine which could solve the halting problem, contradicting Part (a). Similar arguments imply that EMPTY and FINITE are undecidable. Let Z' be the Turing machine whose program is $P = p_1 : * p_2$. Now EQDMN must also be undecidable, since

$$\text{HALT}(Z, x) \Leftrightarrow \text{EQDMN}(Z^x, Z'). \qquad \blacksquare$$

In order to prove Parts (c) through (g) more formally, the above proof must be expressed in terms of Gödel words of Turing machines; it will also be necessary to show that $gw(Z^x)$ can be obtained from $gw(Z)$ and x by recursive means. (This may be intuitively clear, but must be established mathematically.)

4. KLEENE'S S_n^m THEOREM

In this section we prove the S_n^m theorem and complete the proof just given. Some of the machinery we set up will be useful in the next chapter (in particular the results relating to substitution).

We know that if $f(\vec{y}_m, \vec{x}_n)$ is an $(m + n)$ary computable function and we fix $\vec{y}_m$ at constant values, the resulting $n$ary function is also computable (since it is an explicit transform of f). The S_n^m theorem says that not only is the $n$ary function computable but that *we can effectively compute the Gödel word of a machine which computes it, given the Gödel word of a machine which computes f, and the values of $y_1, \ldots, y_m$.*

DEFINITION V.4.1

SUBST(x, u, v) is defined and equal to y iff x contains no overlapping occurrences of u, and y is the result of replacing every occurrence of u in x by v.

LEMMA V.4.2

SUBST is a recursive function with a recursive domain.

PROOF

The condition that x contain no overlapping occurrences of u is S-rudimentary and so is recursive as follows:

$$\text{NOV}(x, u) \Leftrightarrow (\forall r, s)_{\mathbf{P}x}[ru\mathbf{B}x \wedge su\mathbf{B}x \wedge r \neq s \Rightarrow (ru\mathbf{B}s \vee su\mathbf{B}r)].$$

The domain of SUBST is recursive by

$$\text{DOMAIN}(x, u, v) \Leftrightarrow \text{NOV}(x, u).$$

If we know that $\text{NOV}(x, u)$ is true, the following informal algorithm will convert the pair (x, λ) to the pair $(\lambda, \text{SUBST}(x, u, v))$.

  (i)   Let $(p, q) = (x, \lambda)$.
  (ii) If u ends p, then construct the pair (r, s) such that $p = ru$ and $s = vq$, and go to (iv).
  (iii) If u does not end p and $p \neq \lambda$, let a be the rightmost symbol of p, and construct the pair (r, s) such that $p = ra$ and $s = aq$.

(iv) If $r \neq \lambda$ then replace p by r, q by s and go to step (ii); otherwise STOP.

For example, if $x = 12112122$, $u = 12$, and $v = 21$, the successive pairs produced are: $(12112122,\ \lambda) \rightarrow (1211212,\ 2) \rightarrow (12112,\ 212) \rightarrow (121,\ 21212) \rightarrow (12,\ 121212) \rightarrow (\lambda,\ 21121212)$. Let $CP(p, q, r, s, u, v)$ be true iff (p, q) and (r, s) are consecutive pairs in such a sequence, and u, v are as above. This is S-rudimentary as follows:

$$CP(p, q, r, s, u, v) \Leftrightarrow (uEp \wedge p = ru \wedge s = vq)$$
$$\vee \left(\neg\, uEp \wedge (\exists a)_{\mathbf{P}_p}(a \in A \wedge aEp \right.$$
$$\left. \wedge\, p = ra \wedge s = aq) \right).$$

We now use the technique of sequence coding to express the fact that (x, λ) can be converted to (λ, y) as follows:

$$SUBST(x, u, v) = y$$
$$\Leftrightarrow NOV(x, u) \wedge \exists w (\exists \Delta, \#)_{\mathbf{P}_w}$$
$$[Seq(w, \Delta) \wedge FIRST(w, \# x \#) \wedge LAST(w, \# \# y)$$
$$\wedge (\forall \alpha, \beta)_{\mathbf{P}_w}[ADJ(\alpha, \beta, w) \Rightarrow Seq(\alpha, \#) \wedge Seq(\beta, \#)$$
$$\wedge (\exists p, q)_{\mathbf{P}_\alpha}(\exists r, s)_{\mathbf{P}_\beta}(\alpha = \# p \# q$$
$$\wedge \beta = \# r \# s \wedge CP(p, q, r, s, u, v))]].$$

Thus the predicate $SUBST(x, u, v) = y$ is recursively enumerable. By Theorem V.2.4 SUBST is a recursive function. ∎

DEFINITION V.4.3

Let $Z_1 = (A, P_1)$ and $Z_2 = (A, P_2)$ be Turing machines with programs $P_1 = p_1 : i_1 \cdots p_k : i_k \quad p_{k+1}$ and $P_2 = q_1 : j_1 \cdots q_l : j_l \quad q_{l+1}$. The **catenation** of Z_1 with Z_2 is the machine $Z_1 Z_2 = (A, P)$, where by definition $P = p_1 : i_1 \cdots p_k : i_k \ P_2'$ and P_2' is obtained from P_2 by replacing every occurrence of any label q_i by p_{i+k}.

The function $TMC(z_1, z_2)$ is defined iff z_1 and z_2 are Gödel words of Turing machines Z_1 and Z_2, and has value $TMC(z_1, z_2) = gw(Z_1 Z_2)$. Clearly $Z_1 Z_2$ is a Turing machine, which behaves as follows: First, $Z_1 Z_2$ will apply Z_1 to its input. If Z_1 halts, Z_2 will then be applied to the tape resulting from the application of Z_1. ∎

THEOREM V.4.4

$TMC(z_1, z_2)$ is a recursive function with a recursive domain.

PROOF

The domain of TMC is recursive by

$$\text{DOMAIN}(z_1, z_2) \Leftrightarrow \text{TM}(z_1) \wedge \text{TM}(z_2).$$

The first label of both $gw(Z_1)$ and $gw(Z_2)$ is 21^62. The first label of $gw(Z_1Z_2)$ is 21^62 and each label $21^{5+i}2$ in $gw(Z_2)$ becomes $21^{5+i+k}2$ in $gw(Z_1Z_2)$. Thus $gw(P_2')$ is merely the result of substituting 21^{5+k} for 21^5 in $gw(P_2)$. The following formula embodies this idea, and shows by Theorem V.2.3 that TMC is recursive:

$$\begin{aligned}\text{TMC}(z_1, z_2) = z \Leftrightarrow (\exists p, q, z_1', z_2')_{\mathbf{P}z} \, [\text{TM}(z_1) \wedge \text{TM}(z_2) \\ \wedge z_1 = z_1'q \wedge \text{Label}(q) \wedge q = p12 \\ \wedge \text{SUBST}(z_2, 21^5, p) = z_2' \wedge z = z_1' \, z_2'].\end{aligned} \quad \blacksquare$$

THEOREM V.4.5   **(The Kleene S_n^m Theorem)**

For each $m, n \geq 1$ there is a recursive function S_n^m such that if

(i) $f(\vec{y}_m, \vec{x}_n)$ is computed by Turing machine Z with Gödel word z, and if
(ii) $(y_1^0, \ldots, y_m^0)$ is a fixed m-uple from $(A^*)^m$,

then $S_n^m(z, y_1^0, \ldots, y_m^0)$ is the Gödel word of a machine which computes the $n$ary function

$$g(\vec{x}_n) = f(y_1^0, \ldots, y_m^0, \vec{x}_n).$$

PROOF

We treat only the case $m = 1$, for notational simplicity. The general case is entirely similar. Let $y = b_1 \cdots b_l$, where each $b_i \in A$. Define the function $W(y)$ to equal $gw(Z_y)$, where the program of Z_y is $p_1 : L$ $p_2 : L$ $p_3 : b_l$ $p_4 : L$ $p_5 : b_{l-1} \cdots p_{2l+1} : b_1$ p_{2l+2}. Clearly Z_y will transform the tape

$$\mathbf{x}_1 * x_2 * \cdots * x_n \quad \text{to} \quad \mathbf{y} * x_1 * \cdots * x_n.$$

It is clear that $Z_y Z$ computes $g(\vec{x}_n)$, so that we can define S_n^1 by

$$S_n^1(z, y) = \text{TMC}(W(y), z).$$

To show that W is recursive we use a similar technique to that of Lemma V.4.2, and construct a sequence which converts the pair word

$\sharp yp_1 : Lp_2$ to $\sharp\sharp W(y)$ a step at a time. Each step will convert a word of the form $\sharp xb\sharp up_i$ (where $x \in A^*$, $b \in A$, and p_1 is a label) into the word of the form $\sharp x\sharp up_{i+1} : L\ p_{i+2} : b_{i+3}$. Of course, this is complicated somewhat by the fact that z is in encoded form.

The following shows that the predicate $W(y) = z$ is recursively enumerable:

$$W(y) = z$$
$$\Leftrightarrow \exists w, \Delta, \sharp[\text{Seq}(w, \Delta) \wedge \text{First}(w, \sharp y\sharp 21^6 2 : L\ 21^7 2)$$
$$\wedge\ \text{Last}(w, \sharp\sharp z) \wedge (\forall \alpha, \beta)_{\mathbf{P}_w}[\text{ADJ}(\alpha, \beta, w)$$
$$\Rightarrow \text{Seq}(\alpha, \sharp) \wedge \text{Seq}(\beta, \sharp) \wedge (\exists p, r, u)_{\mathbf{P}_\beta}(\exists x, b)_{\mathbf{P}_\alpha}(\alpha = \sharp xb\sharp up$$
$$\wedge\ \text{Label}(p) \wedge p = 2r2 \wedge b \in A$$
$$\wedge\ \beta = \sharp x\sharp up : L\ 2r12 : b\ 2r112)]].$$

Since W is total it is recursive, so that S_n^1 is also recursive. ∎

We now give a more formal proof for parts (d) through (g) of Theorem V.3.5. We construct a recursive function f such that for all Turing machines Z and inputs $x \in A^*$,

$$f(gw(Z), x) = gw(Z^x).$$

Let $Z'' = (A, P'')$, where $A = \{a_1, \ldots, a_m\}$ and

$$P'' = p_1 : * p_2 : R\quad p_3 : Ja_1 p_1 \cdots p_{m+2} : Ja_m p_1\quad p_{m+3}.$$

Clearly the Z^x of Theorem V.3.5 can be defined to equal $Z''Z_x Z$, where Z_x is as in the previous theorem. Now $f(z, x)$ can be defined by

$$f(z, x) = \text{TMC}(gw(Z''), \text{TMC}(W(x), z)).$$

Clearly f is recursive. The predicates EMPTY, ALL, FINITE, and EQDMN cannot be decidable, because of the following:

$$\text{HALT}(z, x) \Leftrightarrow \neg\ \text{EMPTY}(f(z, x)),$$
$$\text{HALT}(z, x) \Leftrightarrow \text{ALL}(f(z, x)),$$
$$\text{HALT}(z, x) \Leftrightarrow \neg\ \text{FINITE}(f(z, x)),$$
$$\text{HALT}(z, x) \Leftrightarrow \text{EQDMN}(f(z, x), \mathbf{p}_1 : * \mathbf{p}_2).$$

Exercises

1. Prove formally that the predicate $\text{HALT}(Z, \lambda)$ is undecidable.
2. Construct formal versions of the problems at the end of Section II.1; and solve the resulting problems.

VI OTHER FORMULATIONS OF COMPUTABILITY

In this chapter we show that two more, quite different formulations of the basic ideas of computability are equivalent to Turing-computability. These are the μ-recursive and general recursive functions of Gödel, Herbrand and Kleene [K1], and the canonical production systems due to Post [P1].

1. DEFINITION OF COMPUTABILITY BY RECURSION

In this section we show that the computable functions can be defined by sets of equations, without reference to predicates. Our formulation is essentially that of Kleene [K1], extended to arbitrary alphabets.

The basic idea is to define a function by means of a finite set of equations which completely determine its value for specific arguments, usually in a recursive or repetitive manner. We give a number of examples, including both informal and formal versions. In all cases $A = \{1, 2\}$. Most of these definitions will be used later.

1. Let $C(x, y) = xy$. A recursive definition of C is:

$$C(x, \lambda) = x \qquad\qquad x\lambda = x$$
$$C(x, y1) = C(x, y)1 \quad \text{or} \quad x(y1) = (xy)1$$
$$C(x, y2) = C(x, y)2 \qquad x(y2) = (xy)2.$$

We could compute $C(12,212)$ recursively as follows:

$$C(12, \lambda) = 12$$
$$C(12, 2) = C(12, \lambda)2 = 122$$
$$C(12, 21) = C(12, 2)1 = 1221$$
$$C(12, 212) = C(12, 21)2 = 12212.$$

2. Let $S(x) = x + 1$ in dyadic notation. Then

$$S(\lambda) = 1 \qquad\qquad \lambda + 1 = 1$$
$$S(x1) = x2 \quad \text{or} \quad x1 + 1 = x2$$
$$S(x2) = S(x)1 \qquad x2 + 1 = (x + 1)1.$$

$S(2122)$ could be computed by

$$S(21) = 22,$$
$$S(212) = S(21)1 = 221,$$
$$S(2122) = S(212)1 = 2211.$$

3. Let $l(x) = |x|$. This is definable in terms of $S(x)$ as follows:

$$l(\lambda) = \lambda \qquad\qquad |\lambda| = \lambda$$
$$l(x1) = S(l(x)) \quad \text{or} \quad |x1| = |x| + 1$$
$$l(x2) = S(l(x)) \qquad |x2| = |x| + 1.$$

Using $S(x)$ it is possible to define in a recursive manner all of the usual functions of number theory. We illustrate with addition, multiplication and exponentiation (letting $0^0 = 1$).

4. $p(x, y) = x + y$

$$p(x, \lambda) = x \qquad\qquad x + \lambda = x$$
$$p(x, S(y)) = S(p(x, y)) \quad \text{or} \quad x + (y + 1) = (x + y) + 1.$$

5. $t(x, y) = x \cdot y$

$$t(x, \lambda) = \lambda \qquad\qquad x \cdot \lambda = \lambda$$
$$t(x, S(y)) = p(t(x, y), x) \quad \text{or} \quad x \cdot (y + 1) = x \cdot y + x.$$

6. $e(x, y) = x^y$

$$e(x, \lambda) = 1$$
$$e(x, S(y)) = r(e(x, y), x)$$
or
$$x^\lambda = 1$$
$$x^{y+1} = x^y \cdot x.$$

We now make explicit the types of equations and sets of equations which are permissible, and how they are used to compute functions.

To begin with let $x, y, z, x_1, x_2, \ldots$ be a list of *variable names* and let $f, g, h, f_1, f_2, \ldots$ be a list of *function names*. Let $A = \{a_1, \ldots, a_m\}$ be an arbitrary alphabet.

DEFINITION VI.1.1

Term and **equation** are defined as follows:

1. λ is a term.
2. Any variable name is a term.
3. If t is a term then ta_i is a term for $i = 1, 2, \ldots, m$.
4. If $n \geq 1$, $t_1, t_2, \ldots, t_n$ are terms and f is a function name then the word: $f(t_1, \ldots, t_n)$ is a term.
5. If s and t are terms then $s = t$ is an equation.
6. No expression is a term or an equation unless it can be shown so by Rules 1 through 5.   ∎

Thus each term or equation is a word over the alphabet $A \cup P \cup V \cup F$, where P is the set consisting of (,), =, and the comma, and V and F are finite sets of variable names and function names, respectively. A word over the alphabet A, that is, not containing any variable or function letters or punctuation symbols, will be called an *A*-**word**. We now define the exact meaning of a computation by a set of equations E. A variable written in boldface (for example, **x**) will denote an *A*-word, rather than one of the variable or function names in $V \cup F$.

DEFINITION VI.1.2

Let E be a finite set of equations. A **computation** or **deduction** by E is any finite sequence $\alpha_1, \alpha_2, \ldots, \alpha_p$ of equations such that for every $i = 1, 2, \ldots, p$, one of the following holds:

1. α_i is an equation of E; or

2. there is an equation α_j with $j < i$ such that α_i is obtained from α_j by replacing *all occurrences* of a variable name u in α_j by an A-word $\mathbf{u}$; or

3. there are equations α_j, α_k with $j, k < i$ such that α_j is of the form $f(\mathbf{u}_1, \ldots, \mathbf{u}_n) = \mathbf{w}$, where $\mathbf{u}_1, \ldots, \mathbf{u}_n$, $\mathbf{w}$ are all A-words, and α_i is obtained from α_k by replacing *one ocurrence* of the word: $f(\mathbf{u}_1, \ldots, \mathbf{u}_n)$ in α_k by $\mathbf{w}$. ∎

For example, the following would be a computation of $S(2122)$ by the equations of Example 2:

$$S(x1) = x2, \ S(21) = 22, \ S(x2) = S(x)1,$$
$$S(212) = S(21)1, \ S(212) = 221, \ S(2122) = S(212)1,$$
$$S(2122) = 2211.$$

The rules applied in order were 1, 2, 1, 2, 3, 2, and 3.

DEFINITION VI.1.3

A partial function $f: (A^*)^n \to A^*$ is **general recursive** iff there is a finite set of equations E with function name f such that

1. if $f(\mathbf{x}_1, \ldots, \mathbf{x}_n) = \mathbf{y}$ where $\mathbf{x}_1, \ldots, \mathbf{x}_n$, $\mathbf{y}$ are A-words, then there is a computation by E which ends with the equation: $f(\mathbf{x}_1, \ldots, \mathbf{x}_n) = \mathbf{y}$; and

2. if E has a computation ending in the equation: $f(\mathbf{x}_1, \ldots, x_n) = \mathbf{y}$ where $\mathbf{x}_1, \ldots, \mathbf{x}_n$, $\mathbf{y}$ are all A-words, then $f(\mathbf{x}_1, \ldots, \mathbf{x}_n)$ is defined and equal to $\mathbf{y}$. ∎

Notice that Condition 2 rules out the possibility that E might lead to contradictory computations. For example E could not contain the two equations $f(1) = 2$ and $f(x) = 1$, since that implies $f(1) = 1$ and $f(1) = 2$. The problem of deciding whether or not E actually defines a function unambiguously can be very difficult in specific cases and is in fact undecidable in the general case.

Exercises

1. Construct sets of equations which define the following functions:

    (a) $d(x, y) = \begin{cases} x - y & \text{if} \quad x \geq y, \\ 0 & \text{if} \quad x < y; \end{cases}$

(b) $sq(x) = \begin{cases} \sqrt{x} & \text{if } x \text{ is a perfect square,} \\ \text{undefined} & \text{otherwise.} \end{cases}$

2. Suppose E is a set of equations which defines a one-to-one function $f: A^* \to A^*$, and let g be the inverse of f, so $g(x) = y$ if $f(y) = x$, and $g(x)$ is undefined if x is not in the range of f. A set of equations defining g may be obtained by adding a single equation to E. What is this equation?

3. Show that the following is undecidable. Given a set of equations E and a function name f, the problem is to determine whether there are A-words $\mathbf{x}$, $\mathbf{y}$, $\mathbf{z}$ such that

   (a) $\mathbf{y} \neq \mathbf{z}$,
   (b) E has deductions which end in "$f(\mathbf{x}) = \mathbf{y}$" and "$f(\mathbf{x}) = \mathbf{z}$".

2. EVERY GENERAL RECURSIVE FUNCTION IS RECURSIVE

Using the techniques developed in Chapter III, it is a relatively straightforward matter to show that every general recursive function is recursive. The auxiliary symbols theorem will be quite useful in this. We first redefine the sequence predicates used so much in Chapters III and V in terms of a separator Δ which is now a new symbol rather than a word of the form $21 \cdots 12$. If A is any alphabet and $\Delta \notin A$, a *sequence word* is simply any word y over $A \cup \{\Delta\}$. The sequence $\alpha_1, \alpha_2, \ldots, \alpha_m$ of words from A^* is represented by $\alpha_1 \Delta \alpha_2 \Delta \cdots \Delta \alpha_m$. Thus every word over $A \cup \{\Delta\}$ represents a sequence (perhaps with $m = 1$), and every sequence can be represented by a unique word over $A \cup \{\Delta\}$; so the predicate $\text{Seq}(y, \Delta)$ is no longer necessary. The following predicates will be useful:

1. $\text{In}(\alpha, y)$ true iff $y = \alpha_1 \Delta \cdots \Delta \alpha_m$ and $\alpha = \alpha_i$ for some i, $1 \leq i \leq m$;
2. $\text{First}(y, \alpha)$, true iff $\alpha = \alpha_1$;
3. $\text{Last}(y, \alpha)$, true iff $\alpha = \alpha_m$;
4. $\text{Adj}(\alpha, \beta, y)$, true iff $\alpha = \alpha_i$ and $\beta = \alpha_{i+1}$ for some i $(1 \leq i < m)$;
5. $\alpha \underset{y}{<} \beta$, true iff for some i, j, $\alpha = \alpha_i$ and $\beta = \alpha_j$, where $1 \leq i < j \leq m$.

LEMMA VI.2.1

All the above are S-rudimentary over $A \cup \{\Delta\}$.

PROOF

1. $\text{In}(\alpha, y) \Leftrightarrow (\alpha = y \vee \alpha\Delta\mathbf{B}y \vee \Delta\alpha\Delta\mathbf{P}y \vee \Delta\alpha\mathbf{E}y) \wedge \neg \Delta\mathbf{P}\alpha,$
2. $\text{First}(y, \alpha) \Leftrightarrow (\alpha = y \vee \alpha\Delta\mathbf{B}y) \wedge \neg \Delta\mathbf{P}\alpha,$
3. $\text{Last}(y, \alpha) \Leftrightarrow (\alpha = y \vee \Delta\alpha\,\mathbf{E}y) \wedge \neg \Delta\mathbf{P}\alpha,$
4. $\text{Adj}(\alpha, \beta, y) \Leftrightarrow (\alpha\Delta\beta = y \vee \alpha\Delta\beta\Delta\mathbf{B}y \vee \Delta\alpha\Delta\beta\Delta\mathbf{P}y \vee \Delta\alpha\Delta\beta\mathbf{E}y)$
   $\wedge \neg \Delta\mathbf{P}\alpha\beta,$
5. $\alpha \underset{y}{<} \beta \Leftrightarrow (\exists u, v)_{\mathbf{P}y} [\text{Last}(u, \alpha) \wedge u\Delta\mathbf{B}y$

   $\wedge \text{Last}(v, \beta) \wedge (v\Delta\mathbf{B}y \vee v = y) \wedge u\mathbf{B}v \wedge u \neq v].$ ∎

Suppose now that E is a set of equations over $A \cup P \cup V \cup F$. As before V and F are the sets of variable and function names which occur in E, and P is the set of punctuation symbols. In order to avoid confusion of the formal symbols $(,), =$, comma of E with those of our formulas we denote the symbols of P by $\prec, \succ, \approx$ (for $=$), and c (for comma). Let $B = A \cup P \cup V \cup F \cup \{\Delta, \#\}$, where $\Delta, \#$ are new symbols. We show that if E defines $f: (A^*)^n \to A^*$, then f is recursive over B. By Theorem IV.4.5, this implies that f is recursive over A.

Redefine $\text{Subst}(x, u, v)$ to be the result of substituting v in place of each occurrence of u in x, assuming that no two occurrences of u overlap in x and that $\#$ is not part of x, u, or v.

LEMMA VI.2.2

Subst is a recursive function over B.

PROOF

Very similar to that of Lemma V.4.2. Let $\text{OK}(x, u, v)$ be true iff no occurrences of u overlap in x and $\#$ is not part of x, u, or v. This is S-rudimentary by

$\text{OK}(x, u, v) \Leftrightarrow \neg \#\mathbf{P}xuv$

$\wedge (\forall r, s)_{\mathbf{P}x} [r u\mathbf{B}x \wedge su\mathbf{B}x \wedge r \neq s \Rightarrow (ru\mathbf{B}s \vee su\mathbf{B}r)].$

Thus the domain of Subst is recursive. Let $CP(p, q, r, s, u, v)$ be exactly as in Lemma V.4.2. We now have the following:

$$\text{Subst}(x, u, v) = y \Leftrightarrow \text{OK}(x, u, v) \wedge \exists w[\text{First}(w, x\sharp) \wedge \text{Last}(w, \sharp y)$$
$$\wedge \, (\forall \alpha, \beta)_{\mathbf{P}w}(\text{Adj}(\alpha, \beta, w)$$
$$\Rightarrow (\exists p, q)_{\mathbf{P}\alpha}(\exists r, s)_{\mathbf{P}\beta}(\alpha = p\sharp q \wedge \beta = r\sharp s$$
$$\wedge \, CP(p, q, r, s, u, v)))].$$

Thus $\text{Subst}(x, u, v) = y$ is recursively enumerable, so by Theorem V.2.4 Subst is a recursive function. ∎

THEOREM VI.2.3

If $f: (A^*)^n \to A^*$ is a general recursive partial function then f is recursive over A.

PROOF

Let f be general recursive. Then there must be a finite set of equations E which defines f. Let p be the maximum number of arguments of any function name in E.

We define the following predicates and function over B by

(i)   Aword(x) is true iff $x \in A^*$.
(ii)   Leftside(t) is true iff t equals the word: $g\dotplus\mathbf{u}_1 c \cdots c\mathbf{u}_k\dotplus$, for some $\mathbf{u}_1, \ldots, \mathbf{u}_k \in A^*$, $g \in F$ and $k \leq p$.
(iii) DEDN(y) is true iff $y = \alpha_1 \Delta \alpha_2 \cdots \Delta \alpha_k$ and $\alpha_1, \ldots, \alpha_k$ is a deduction by E.
(iv)   CMP$(x_1, \ldots, x_n, y)$ is true iff $x_1, \ldots, x_n$ are A-words, DEDN(y) is true as above, and α_k equals the word: $f\dotplus x_1 c \cdots c x_n\dotplus \approx \mathbf{w}$ for some $\mathbf{w} \in A^*$.
(v) $L(y) = w$ iff y is as above, w is an A-word, and α_k ends in $\approx w$; $L(y)$ is undefined otherwise.

It is immediate from these definitions that for all $x_1, \ldots, x_n \in A^*$ either

$$f(x_1, \ldots, x_n) = L(\mu y \text{CMP}(x_1, \ldots, x_n, y)) \tag{1}$$

or both sides of this equation are undefined.

We now show that each of these is S-rudimentary. These formulas use the fact that every finite predicate is S-rudimentary:

$$\text{Aword}(x) \Leftrightarrow (\forall b)_{\mathbf{P}x}(b \in B \Rightarrow b \in A),$$

$$\text{Leftside}(t) \Leftrightarrow (\exists u_1, \ldots, u_p, g)_{\mathbf{P}t}$$

$$[\text{Aword}(u_1) \wedge \cdots \wedge \text{Aword}(u_p) \wedge g \in F$$

$$\wedge\ (t = g \,\overline{+}u_1\overline{\mathbf{+}}\ \vee\ t = g\overline{+}u_1cu_2\overline{\mathbf{+}}\ \vee\ \cdots\ \vee\ t = g\overline{+}u_1c \cdots cu_p\overline{\mathbf{+}})].$$

To show that DEDN(y) is S-rudimentary we exhibit a formula that parallels Definition VI.1.2.

$$\text{DEDN}(y) \Leftrightarrow (\forall\alpha)_{\mathbf{P}y}[\text{In}(\alpha, y) \Rightarrow \{\alpha \in E$$

$$\vee\ (\exists\beta, u, \mathbf{u})_{\mathbf{P}y}(\beta \underset{y}{<} \alpha \wedge u \in V \wedge \text{Aword}(\mathbf{u})$$

$$\wedge\ \text{Subst}(\beta, u, \mathbf{u}) = \alpha)$$

$$\vee\ (\exists\beta, \gamma, t, w, u, v)_{\mathbf{P}y}(\beta \underset{y}{<} \alpha \wedge \gamma \underset{y}{<} \alpha$$

$$\wedge\ \text{Leftside}(t) \wedge \text{Aword}(w)$$

$$\wedge\ \gamma = t \approx w \wedge \beta = utv \wedge \alpha = uwy)\}].$$

We now see that CMP and L are S-rudimentary:

$$\text{CMP}(x_1, \ldots, x_n, y) \Leftrightarrow \text{DEDN}(y)$$

$$\wedge\ \text{Aword}(x_1) \wedge \cdots \wedge \text{Aword}(x_n)$$

$$\wedge\ (\exists w, z)_{\mathbf{P}y}[\text{Last}(y, z)$$

$$\wedge\ z = f\overline{+}x_1c \cdots cx_n\overline{\mathbf{+}} \approx w \wedge \text{Aword}(w)];$$

$$L(y) = w \Leftrightarrow \text{Aword}(w) \wedge\ \approx w\mathbf{E}y.$$

Since CMP and L are S-rudimentary, by formula (1) f is μ-rudimentary over B and so is recursive over B. By Theorem IV.4.5 f is recursive over A. ∎

3. EVERY RECURSIVE FUNCTION IS GENERAL RECURSIVE

We prove the result of the title above, for alphabets with two or more letters. First we show that the total characteristic function of any S-rudimentary predicate is general recursive.

DEFINITION VI.3.1

A predicate $R(\vec{x}_n)$ is **GR-decidable** iff its total characteristic function $C_R^t(\vec{x}_n)$ is general recursive.

LEMMA VI.3.2

Let $A = \{a_1, a_2, \ldots, a_m\}$. Each function (predicate) in the following list is general recursive (GR-decidable) over A:

1. $x = y$
2. $C(x, y)$ (equals xy)
3. $xy = z$
4. not(x) (equals λ if $x \neq \lambda$, a_1 if $x = \lambda$)
5. and(x, y) (equals λ if $x = \lambda$ and $y = \lambda$, a_1 otherwise)
6. or(x, y) (equals λ if $x = \lambda$ or $y = \lambda$, a_1 otherwise)
7. $S(x)$ ($= x + 1$ in $m$adic notation).

PROOF

The following set of equations defines all of these functions; "Eq" and "Cat" denote the total characteristic functions of $x = y$ and $xy = z$, respectively.

1. $\text{Eq}(\lambda, \lambda) = \lambda$
    $\text{Eq}(xa_i, ya_i) = \text{Eq}(x, y)$ $(i = 1, \ldots, m)$
    $\text{Eq}(xa_i, ya_j) = a_1$ $(i, j = 1, \ldots, m, i \neq j)$
    $\text{Eq}(xa_i, \lambda) = a_1$ $(i = 1, \ldots, m)$
    $\text{Eq}(\lambda, ya_i) = a_1$ $(i = 1, \ldots, m)$
2. $C(x, \lambda) = x$
    $C(x, ya_i) = C(x, y)a_i$ $(i = 1, \ldots, m)$
3. $\text{Cat}(x, y, z) = \text{Eq}(C(x, y), z)$
4. $\text{not}(\lambda) = a_i$
    $\text{not}(xa_i) = \lambda$ $(i = 1, \ldots, m)$
5. $\text{and}(x, y) = \text{Cat}(x, y, \lambda)$
6. $\text{or}(x, y) = \text{not}(\text{and}(\text{not}(x), \text{not}(y)))$
7. $S(\lambda) = a_1$
    $S(xa_i) = xa_{i+1}$ $(i = 1, \ldots, m - 1)$
    $S(xa_m) = S(x)a_1$.   ∎

LEMMA VI.3.3

If $R(\vec{x}_n)$ and $S(\vec{x}_n)$ are GR-decidable, then so are $\neg R(\vec{x}_n)$, any explicit transform $R(\vec{\xi}_n)$ of $R(\vec{x}_n)$, $R(\vec{x}_n) \wedge S(\vec{x}_n)$, and $R(\vec{x}_n) \vee S(\vec{x}_n)$.

PROOF

Let the characteristic functions of the predicates just named be denoted by $K_r(\vec{x}_n)$, $K_s(\vec{x}_n)$, $K_\neg(\vec{x}_n)$, $K_\xi(\vec{y}_m)$, $K_\wedge(\vec{x}_n)$, and $K_\vee(\vec{x}_n)$, respectively. The following set of equations showing that these are GR-decidable:

$$K_\neg(\vec{x}_n) = \text{not}(K_r(\vec{x}_n)),$$
$$K_\xi(\vec{y}_m) = K_r(\vec{\xi}_n),$$
$$K_\wedge(\vec{x}_n) = \text{and}(K_r(\vec{x}_n), K_s(\vec{x}_n)),$$
$$K_\vee(\vec{x}_n) = \text{or}(K_r(\vec{x}_n), K_s(\vec{x}_n)). \qquad \blacksquare$$

LEMMA VI.3.4

If $R(\vec{x}_n, y)$ is GR-decidable, then so are $(\exists z)_{\leq y} R(\vec{x}_n, z)$ and $(\forall z)_{\leq y} R(\vec{x}_n, z)$.

PROOF

Denote the total characteristic functions by K_r, $K_\exists$, and $K_\forall$, respectively. The following equations define $K_\exists$ and $K_\forall$:

$$K_\exists(\vec{x}_n, \lambda) = K_r(\vec{x}_n, \lambda),$$
$$K_\exists(\vec{x}_n, S(y)) = \text{or}(K_\exists(\vec{x}_n, y), K_r(\vec{x}_n, S(y)))$$
$$K_\forall(\vec{x}_n, \lambda) = K_r(\vec{x}_n, \lambda),$$
$$K_\forall(\vec{x}_n, S(y)) = \text{and}(K_\forall(\vec{x}_n, y), K_r(\vec{x}_n, S(y))). \qquad \blacksquare$$

LEMMA VI.3.5

Every S-rudimentary predicate is GR-decidable.

PROOF

As a result of the preceding lemmas it is only necessary to show that if $R(\vec{x}_n, y)$ is GR-decidable, then $(\exists z)_{\mathbf{P}y} R(\vec{x}_n, z)$ and $(\forall z)_{\mathbf{P}y} R(\vec{x}_n, z)$ are GR-decidable. This, however, is immediate by the following:

$$x\mathbf{P}y \Leftrightarrow (\exists u, v, w)_{\leq y}(ux = v \wedge vw = y)$$
$$(\exists z)_{\mathbf{P}y} R(\vec{x}_n, z) \Leftrightarrow (\exists z)_{\leq y}(z\mathbf{P}y \wedge R(\vec{x}_n, z))$$
$$(\forall z)_{\mathbf{P}y} R(x_n, z) \Leftrightarrow (\forall z)_{\leq y}(z\mathbf{P}y \Rightarrow R(\vec{x}_n, z)). \qquad \blacksquare$$

LEMMA VI.3.6

If $R(\vec{x}_n, y)$ is a GR-decidable predicate and

$$f(\vec{x}_n) = \mu y \, R(\vec{x}_n, y),$$

then f is general recursive.

PROOF

Let E_r be a set of equations which defines K_r, the total characteristic function of R. Without loss of generality E_r contains a definition of the successor function $S(x) = x + 1$, and does not contain function names f, g, or h.

We claim that the following set of equations E defines $f(\vec{x}_n)$:

1. the equations of E_r,
2. $h(\vec{x}_n, y, \lambda) = y$,
    $h(\vec{x}_n, y, a_1) = g(\vec{x}_n, S(y))$,
3. $g(\vec{x}_n, y) = h(\vec{x}_n, y, K_r(\vec{x}_n, y))$,
4. $f(\vec{x}_n) = g(\vec{x}_n, \lambda)$.

Now suppose that $\vec{x}_n$ are fixed, and that $y_1, y_2, y_3, \ldots$ is the enumeration in increasing order of the set of all ys in A^* such that $R(\vec{x}_n, y)$ is true. Examination of Equations 2 and 3 reveals that

$$g(\vec{x}_n, y) = \begin{cases} y_1 & \text{if} \quad y \leq y_1, \\ y_2 & \text{if} \quad y_1 < y \leq y_2, \\ y_3 & \text{if} \quad y_2 < y \leq y_3, \\ \vdots \end{cases}$$

Further $g(\vec{x}_n, y)$ will be undefined just in case $R(\vec{x}_n, z)$ is false for all $z \geq y$. Consequently for all $x_1, \ldots, x_n \in A^*$,

$$f(\vec{x}_n) = g(\vec{x}_n, \lambda).$$

To see that E is not self-contradictory, observe that

(a) E_r is not self-contradictory, since it defines K_r; and
(b)   Equations 2 and 3 define g and h in a completely unambiguous way;
(c)   consequently f is defined unambiguously.   ∎

THEOREM VI.3.7

If A contains at least two symbols, then every partial function over A which is recursive is also general recursive.

PROOF

Let $f(\vec{x}_n)$ be recursive. By Theorem V.1.2 there is a word $z_0 \in A^*$ such that for all $\vec{x}_n \in A^*$,

$$f(\vec{x}_n) = U(\mu y \ T_n(z_0, \vec{x}_n, y)),$$

where $T_n(z, x_n, y)$ and $U(y) = w$ are S-rudimentary predicates. By the previous lemmas both are GR-decidable; in addition U is a general recursive function by the last lemma, since for all $y \in A^*$

$$U(y) = \mu z(U(y) = z).$$

Thus there are sets of equations E_u and E_t which define U and the function $t(\vec{x}_n) = \mu y T_n(z_0, \vec{x}_n, y)$. Without loss of generality E_u and E_t contain neither f nor any common function names. Thus the following set of equations defines f:

1. E_u,
2. E_t,
3. $f(\vec{x}_n) = U(t(\vec{x}_n))$.

Thus f is general recursive. ∎

4. THE ONE-LETTER CASE

The original formulation of general recursive functions (for example, [K1]) made no reference to words or alphabets; rather it dealt only with natural numbers. Of course the number n can be identified with the word 1^n, and in fact our Definitions VI.1.1 and VI.1.2 closely resemble that of [K1] if this notational change is made. From this viewpoint Theorem VI.3.7 is unsatisfactory, since it omits the case in which A contains only a single letter. However this can be remedied, as follows:

We only need to consider the difference between a two-letter alphabet $B = \{1, 2\}$, and the one-letter alphabet $A = \{1\}$. The technique to be used is to represent a string $x \in B^*$ by 1^m, where m is the value of x if read as a dyadic integer. Two auxiliary functions "doub" and "ext" on A^* will be needed, such that for all $m, n \geq 0$, $\text{doub}(1^m) = 1^{2m}$ and $\text{ext}(1^m) = 1^n$ if m is the number whose dyadic representation is 1^n, that is, n occurrences of the symbol 1, and is undefined otherwise. Consequently $m = 2^0 + 2^1 + \cdots + 2^{n-1} = 2^n - 1$, where 2^x denotes exponentiation, so $\text{ext}(1) = 1$, $\text{ext}(1^3) = 11$, $\text{ext}(1^7) = 111$, and so forth.

LEMMA VI.4.1

doub and ext are general recursive over $A = \{1\}$.

PROOF

The following equation set defines them:

1. $\text{doub}(\lambda) = \lambda$, $\text{doub}(x1) = \text{doub}(x)11$.
2. $\text{ext}(\lambda) = \lambda$, $\text{ext}(\text{doub}(x)1) = \text{ext}(x)1$.   ∎

LEMMA VI.4.2

Let $A = \{1\}$, $B = \{1, 2\}$, and $f : (A^*)^n \to A^*$, and suppose that $g : (B^*)^n \to B^*$ satisfies: $g(x_n) = f(x_n)$ if $x_1, \ldots, x_n \in A^*$. Then f is general recursive over A if g is general recursive over B.

PROOF

Let E be a set of equations over B which defines g, and which does not contain the name f. We shall construct a set E_f over A which defines f, by modifying each equation in E; in particular the encoding technique mentioned above for mapping B^* into A^* will be applied to each equation e of E, yielding an equation e_A in E_f which specifies the same relation as e does, but in encoded terms.

Recalling Definition VI.1.1, e_A is defined inductively as follows:

1.   If term t is λ, then $t_A = \lambda$.
2.   If t is a variable name, then $t_A = t$.
3a. If $t = u1$, where u is a term, then $t_A = \text{doub}(u_A)1$.
3b. If $t = u2$, where u is a term, then $t_A = \text{doub}(u_A)11$.

4. If $t = h(u_1, \ldots, u_n)$, where h is a function name and $u_1, \ldots, u_n$ are terms, then $t_A = h(u_{1A}, \ldots, u_{nA})$.
   If "$s = t$" is an equation e, then e_A is "$s_A = t_A$."

E_f is now defined to be

A. e_A for each equation e in E,
B. the equations defining doub and ext,
C. $f(\vec{x}_n) = \text{ext}(g(\vec{x}_n))$.

Following is the reason for statements 3a and 3b: Let $\mathbf{u} \in B^*$, and let its dyadic value be m. Then the dyadic values of $\mathbf{u}1$ and $\mathbf{u}2$ will be $2m + 1$ and $2m + 2$, respectively.

It is easily verified that for all $\vec{x}_n \in A^*$, Equations A and B of E_f will yield a deduction ending in "$g(\vec{x}_n) = 1^m$" if and only if m is the value of $g(\vec{x}_n)$ when read as a dyadic integer. In case $g(\vec{x}_n)$ equals a string consisting of p ones, then m will equal $2^p - 1$, so Equation C will cause a deduction of "$f(\vec{x}_n) = 1^p$," as required. ∎

The following theorem is similar to Theorem VI.3.7, but slightly stronger since it applies to arbitrary alphabets.

THEOREM VI.4.3

If A is any alphabet, a function over A is recursive if and only if it is general recursive.

PROOF

The "if" part is Theorem VI.2.3 (which works for any alphabet A, by the auxiliary symbols theorem). The "only if" part is given by Theorem VI.3.7 for $\sharp A \geq 2$.

If $\sharp A = 1$ and $f: (A^*)^n \to A^*$ is recursive over A, there must be a Turing machine $Z = (A, P)$ which computes it. Construct $Z' = (B, P)$ where $B = \{1, 2\}$; that is, Z' is identical to Z, except that it has alphabet B. Denote the $n$ary function which Z' computes by $g(\vec{x}_n)$. Then by Theorem VI.3.7, g is general recursive over B, that is, there is a set of equations over B which defines g.

Clearly g satisfies the conditions of Lemma VI.4.2, so f is general recursive over A, as required. ∎

5. SEMI-THUE SYSTEMS

In the previous sections we saw that a very powerful method for definition of functions, that is, the use of sets of recursive equations, was exactly equivalent to Turing computability. That equivalence was perhaps surprising *a priori*, because of the comparative simplicity of Turing machines.

We now prove a result which is surprising for the opposite reason; specifically, we introduce the semi-Thue systems, a class of systems which is even simpler than the Turing machines, and show that they in fact can define all and only the recursively enumerable sets.

Informally, a semi-Thue system consists primarily of an *axiom*, which is a word α over an alphabet B, and a finite set of *productions* of the form $x \to y$ where x and y are words over B. Each production is interpreted as a *rewriting rule*, which states that any word of the form uxv can be rewritten in one step to become uyv.

For example, let $B = \{1, 2, 3\}$, let the axiom be 2322, and let the productions be

$$3 \to 132 \quad \text{and} \quad 3 \to 12.$$

Then one can obtain by repeatedly applying these productions, the sequences: 2322, 21222; 2322, 213222, 2112222; or 2322, 213222, 21132222, 211122222. Clearly all words which can be obtained from the axiom and which do not contain 3 must be of the form: 21^n2^n22, where n varies over 1, 2, 3, 4, ...

DEFINITION VI.5.1

Let A and B be alphabets such that $A \subseteq B$

1. A **Semi-Thue production** is any object of the form $x \to y$, where $x, y \in B^*$.
2. A **Semi-Thue system** is a quadruple $S = (A, B, \Pi, \alpha)$ such that A and B are alphabets, Π is a finite set of semi-Thue productions, and α (called the **axiom** of S) is a word over B.

3. A **deduction** by S is a sequence of words $\alpha_1, \ldots, \alpha_p$ over B such that

   (a) α_1 is the axiom α of S; and
   (b) for each i $(1 < i \leq p)$ there is a production $x \to y$ in Π and words u, v over B such that $\alpha_{i-1} = uxv$ and $\alpha_i = uyv$.

4. The **set generated by S** is $G(S) = \{x \,|\, x \in A^*$ and there is a deduction $\alpha_1, \ldots, \alpha_p$ for which $x = \alpha_p\}$. ∎

Thus the informal example above is formally given by

$$S = (\{1, 2\}, \{1, 2, 3\}, \{3 \to 132, 3 \to 12\}, 2322).$$

Further,

$$G(S) = \{21^n 2^{n+2} \,|\, n \geq 1\}.$$

We now show that the sets generated by semi-Thue sets are exactly the recursively enumerable sets.

LEMMA VI.5.2

If $S = (A, B, \Pi, \alpha)$ is a semi-Thue system, then $G(S)$ is a recursively enumerable set.

PROOF

This proof closely parallels that of Theorem VI.2.3. We shall show that $G(S)$ is recursively enumerable over $B \cup \{\Delta\}$, where Δ is a sequence marker as in Section VI.2. By the Auxiliary Symbols Theorem this will imply that $G(S)$ is recursively enumerable over A.

Define $PR(x, y)$ to be true iff $x \to y$ is a production of Π; this is finite and so is recursive (in fact S-rudimentary).

The following formula shows that the predicate "$x \in A^*$" is S-rudimentary:

$$x \in A^* \Leftrightarrow (\forall b)_{\mathbf{P}x}(b \in B \Rightarrow b \in A).$$

Consider the following formula:

$$R(w) \Leftrightarrow w \in A^* \wedge \exists z \{\mathrm{First}(z, \alpha) \wedge \mathrm{Last}(z, w)$$
$$\wedge \,(\forall r, s)_{\mathbf{P}z}[\mathrm{Adj}(r, s, z)$$
$$\Rightarrow (\exists u, v, x, y)_{\mathbf{P}z}(PR(x, y) \wedge r = uxv \wedge s = uyv)]\}.$$

It is easily verified that $R(w)$ is true if and only if $w \in G(S)$. In addition, the part of the formula enclosed in braces { and } is S-rudimentary and so is recursive; thus R is recursively enumerable by Corollary V.1.3 and Theorem V.2.2.   ∎

To show the converse, we first show that semi-Thue systems can mimic the behavior of Turing machines.

LEMMA VI.5.3

Let $Z = (A, P)$ be a Turing machine, let $P = p_1 : i_1 \cdots p_k : i_k \; p_{k+1}$, and let $B = A \cup \{p_1, \ldots, p_{k+1}, *, \Delta\}$. Then there is a semi-Thue system $S = (A, B, \Pi, \lambda)$ such that for any I.D.s α and β of Z, $\alpha \vdash_Z \beta$ iff there is a production $x \to y$ in Π such that $\Delta\alpha\Delta = uxv$ and $\Delta\beta\Delta = uyv$ for some words $u, v \in B^*$.

PROOF

Construction of Π closely parallels Definition III.2.3. For each value of $j = 1, 2, \ldots, k$ suppose P contains $p_j : i$ where $i = i_j$. Then Π is defined as follows, where A_* is an abbreviation for $A \cup \{*\}$:·

   1.   If $i \in A_*$, then Π contains

$$p_j a \to p_j i \quad \text{for each} \quad a \in A_*.$$

   2.   If $i = Jcp_l$, then Π contains

$$p_j c \to p_l c$$

and

$$p_j a \to p_{j+1} a \quad \text{for each} \quad a \in A_* \backslash \{c\}.$$

   3.   If $i = R$, then Π contains

$$p_j ab \to ap_{j+1} b \quad \text{for each} \quad a, b \in A_*,$$
$$p_j a\Delta \to ap_{j+1} * \Delta \quad \text{for each} \quad a \in A_*.$$

   4.   If $i = L$, then Π contains

$$bp_j a \to p_{j+1} ba \quad \text{for each} \quad a, b \in A_*,$$
$$\Delta p_j a \to \Delta p_{j+1} * a \quad \text{for each} \quad a \in A_*.$$

Verification that S behaves as required is straightforward.   ∎

LEMMA VI.5.4

If $T \subseteq A^*$ is a recursively enumerable set, there is a semi-Thue system $S = (A, B, \Pi, \alpha)$ such that $T = G(S)$.

PROOF

The case $T = \varnothing$ is immediate, so suppose $T \neq \varnothing$. Then T is the range of some total recursive function $f: A^* \to A^*$ by Theorem V.2.6. Let $Z = (A, P)$ be a Turing machine which standard computes f, with initial label p and final label q and let $S_0 = (A, B_0, \Pi_0, \lambda)$ be the corresponding semi-Thue system of the previous lemma. Let $B = B_0 \cup \{\#\}$, where $\#$ is a new symbol. We define S to equal $(A, B, \Pi, \#\Delta)$, where Π contains the productions of Π_0, plus

1. $\# \to \#a$ for each $a \in A$,
2. $\# \to \Delta p$,
3. $\Delta qa \to a\Delta q$ for each $a \in A$,
4. $\Delta q * \to \Delta q$,
5. $\Delta q\Delta \to \lambda$.

We claim that $T = G(S)$. Clearly any deduction of a word in A^* by S must proceed in the order:

(a) First, 0 or more applications of productions of Type 1 above, followed by one application of $\# \to \Delta p$, yielding a word of the form $\Delta px\Delta$, where $x \in A^*$; then
(b) a series of productions from Π_0, which simulate Z and so must yield the word $\Delta qf(x)\Delta$ [just in case $f(x)$ is defined]; and finally
(c) a word in A^* can be derived only by applying productions of Types 3, 4, and 5 above, yielding $f(x)$.

Thus $T = \text{Range}(f) \supseteq G(S)$. It is evident that $T \subseteq G(S)$, so $T = G(S)$. ∎

THEOREM VI.5.5

Let $T \subseteq A^*$. Then T is recursively enumerable if and only if $T = G(S)$ for some semi-Thue system.

PROOF

Immediate from Lemmas VI.5.2 and VI.5.4. ∎

6. POST CANONICAL SYSTEMS

The final formulation of computability which we shall consider is the class of *Post canonical systems*, also due to Emil Post [P1]. These systems are of particular interest because they include, as special cases, nearly every form of symbol manipulation system which has been devised. A canonical system contains an *axiom* and a finite set of *productions*, just as for the case of semi-Thue systems; in addition it contains *production variables* which represent arbitrary words. The example of a semi-Thue system which was given in Section VI.5 would correspond to a canonical system whose axiom is 2322, and whose productions are

$$X3Y \rightarrow X132Y \quad \text{and} \quad X3Y \rightarrow X12Y.$$

The production $X3Y \rightarrow X12Y$ which is of the form: an initial word (which we denote by x), followed by the symbol 3, followed by a final word (which we denote by y); can yield another word obtained by catenating x, then 1, then 2, and then y. Another legitimate production would be $X3Y \rightarrow 1Y2X$, which would have the effect of swapping ends of a word, replacing an occurrence of 3 by 2, and prefixing the result with 1.

A canonical production may also have *multiple premises*; an example of such a production might be:

$$X12Y, \ X3Y \rightarrow 1X2Y.$$

This asserts that if we have generated two words, one of the form $x12y$ and the other of the form $x3y$ (for the same x and y), then we may obtain the word $1x2y$.

As in the case of semi-Thue systems, a canonical system P involves two alphabets A, B such that $A \subseteq B$. The set generated consists of only those words $x \in A^*$ which may be derived from the axiom; thus symbols in $B - A$ may be considered as auxiliary symbols (distinct, however, from the production variables). Following is a slightly more complex example, with auxiliary symbols t, u, and v, and production variables X and Y. The axiom is the word u; the productions are as follow:

(a) $u \to t$
(b) $u \to v2$
(c) $tX \to t1X$
(d) $vX2Y, tY \to vX2Y2Y11$
(e) $vX \to X$

In this example, t, u, and v correspond roughly to predicate names. The words of the form tx which are derivable are t (since $u \to t$), $t1$ [by applying production (c) with $X = \lambda$], $t11$ [by applying (c) with $X = 1$, since $t1$ has been deduced], and so forth. Clearly a word tx is deducible iff x is a tally. Table VI.1 shows a deduction of two words over $A = \{1, 2\}$.

TABLE VI.1

Step	Word Deduced	Justification
1	u	axiom
2	$v2$	(b) and line 1
3	2	(e) and line 2 with $X = \lambda$
4	t	(a) and line 1
5	$v2211$	(d) and lines 2 and 4, with $X = Y = \lambda$
6	2211	(e) and line 5

Examination of the productions reveals that this canonical system generates the set of all words of the form $21^0 21^2 21^4 \cdots 21^{2n}$ such that $n = 0, 1, 2, \ldots$.

Using methods of this nature, it is possible to construct canonical systems which closely model a great variety of symbol manipulation systems, including semi-Thue systems, Turing machines, many other mathematical machines, and rules of formula construction and proofs in all the customary axiom systems of formal mathematics. For examples, see Rosenbloom [R2].

Following is a precise definition of Post canonical system. Suppose $V = \{v_1, \ldots, v_m\}$ is an alphabet and let $x, u_1, \ldots, u_m$ be words over some larger alphabet. Then $SB(x, u_m)$ denotes the result of simultaneously replacing every occurrence of v_i in the word x by u_i, for $i = 1, 2, \ldots, m$.

DEFINITION VI.6.1

Let V, A, and B be alphabets such that $A \subseteq B$ and $B \cap V = \emptyset$. V is called a set of **production variables**.

1. A **canonical production** is any object of the form $x_1, \ldots, x_n \to y$, where $x_1, \ldots, x_n$ and y are words over $V \cup B$, such that any symbol $v \in V$ which occurs in y also occurs in at least one of $x_1, \ldots, x_n$.
2. A **Post canonical system** is a quintuple $P = (V, A, B, \Pi, \alpha)$ such that V, A, and B are as above, Π is a finite set of canonical productions, and α (called the **axiom** of P) is a word over B.
3. Let $V = \{v_1, \ldots, v_m\}$. A **deduction** is a sequence $\alpha_1, \alpha_2, \ldots, \alpha_p$ of words over B such that
   (a) α_1 is the axiom α, and
   (b) for each i $(1 < i \le p)$ there exists a production $x_1, \ldots, x_n \to y$ in Π, words $u_1, \ldots, u_m \in B^*$ and words $\alpha_{i_1}, \ldots, \alpha_{i_n}$ $(1 \le i_j < i$ for $j = 1, \ldots, n)$ such that $\alpha_{i_1} = SB(x_1, \vec{u}_m), \ldots,$ $\alpha_{i_n} = SB(x_n, \vec{u}_m)$ and $\alpha_i = SB(y, \vec{u}_m)$.
4. The **set generated by** P (or **the set of theorems of** P) is
   $G(P) = \{x \mid x \in A^*$ and there is a deduction $\alpha_1, \ldots, \alpha_p$ for which $x = \alpha_p\}$ ∎

Thus the example given before this definition is formally given by

$$S = (\{X, Y\}, \{1, 2\}, \{1, 2, u, t, v\}, \Pi, u),$$

where

$$\Pi = \{u \to t, u \to v2, tX \to t1X, vX2Y, tY \to vX2Y2Y11, vX \to X\}.$$

The deduction in the example would be $\alpha_1, \ldots, \alpha_6 = u, v2, 2, t, v2211,$ 2211. The step yielding $\alpha_5 = v2211$ is obtained by $\alpha_2 = v2 = SB(vX2Y, \lambda, \lambda)$, and $\alpha_5 = v2211 = SB(vX2Y211, \lambda, \lambda)$.

LEMMA VI.6.2

If $S = (A, B, \Pi, \alpha)$ is a semi-Thue system, then there is a Post canonical system $P = (\{X, Y\}, A, B, \Pi_1, \alpha)$ such that $G(S) = G(P)$.

PROOF

Define

$$\Pi_1 = \{Xr Y \to Xs Y \,|\, r \to s \text{ is in } \Pi\}.$$

It is easily seen that a sequence $\alpha_1, \ldots, \alpha_p$ of words over B is a deduction by S if and only if it is a deduction by P; consequently $G(S) = G(P)$. ∎

COROLLARY VI.6.3

If $L \subseteq A^*$ is recursively enumerable, then $L = G(P)$ for some Post canonical system P.

PROOF

Immediate from the previous lemma and Theorem VI.5.5. ∎

LEMMA VI.6.4

If $P = (V, A, B, \Pi, \alpha)$ is a Post canonical system, then $G(P)$ is recursively enumerable.

PROOF

This proof is similar to that of Theorems VI.2.3 and VI.5.2. We actually show that $G(P)$ is recursively enumerable over $V \cup B \cup \{\Delta\}$, where Δ is a new symbol not in $V \cup B$.

Let $\pi = x_1, \ldots, x_n \to y$ be a production in Π. Define $Y_\pi(w)$ to be true iff w is of the form $w = \alpha_1 \Delta \cdots \Delta \alpha_p$, where each $\alpha_i \in B^*$, and α_p is obtained from earlier α_is by applying production π as in part 3 of Definition VI.6.1. The relation $\beta = SB(x, \bar{u}_m)$, whose variables range over $V \cup B \cup \{\Delta\}$, can be shown to be recursive in a manner very similar to that of Lemma VI.2.2. Thus the following formula shows that Y_π is recursive.

$$Y_\pi(w) \Leftrightarrow (\exists \beta, \alpha'_1, \ldots, \alpha'_n, u_1, \ldots, u_m)_{\mathbf{P}w}$$

$$\{\text{Last}(w, \beta) \wedge \alpha'_1 \underset{w}{<} \beta \wedge \cdots \wedge \alpha'_n \underset{w}{<} \beta \wedge u_1 u_2 \cdots u_m \in B^*$$

$$\wedge \; \alpha'_1 = SB(x_1, \vec{u}_m) \wedge \cdots \wedge \alpha'_n = SB(x_n, \vec{u}_m)$$

$$\wedge \; \beta = SB(y, \vec{u}_m)\}.$$

Now define $Y(w)$ to be true iff w is of the form $\alpha_1 \Delta \cdots \Delta \alpha_p$, and w is the axiom α, or α_p follows from earlier α_is by applying a single production. Let $\Pi = \{\pi_1, \ldots, \pi_q\}$. Then Y is seen to be recursive by

$$Y(w) \Leftrightarrow \neg \, \Delta \mathbf{P}w \wedge [w = \alpha \vee Y_{\pi_1}(w) \vee \cdots \vee Y_{\pi_q}(w)].$$

It is easily verified that the following formula defines a predicate T over $V \cup B \cup \{\Delta\}$, such that $T(z)$ is true iff z is in $G(P)$:

$$T(z) \Leftrightarrow z \in A^* \wedge \exists y \{(z = y \vee \Delta z \mathbf{E}y) \wedge (\forall w)_{\mathbf{P}y}[(w \Delta \mathbf{B}y \vee w = y) \Rightarrow Y(w)]\}.$$

By Corollary V.1.3 and Theorem V.2.2, T and so $G(P)$ is recursively enumerable over $V \cup B \cup \{\Delta\}$; and so over A.   ∎

THEOREM VI.6.5

A set $L \subseteq A^*$ is generated by a Post canonical system if and only if it is recursively enumerable.

PROOF

Immediate from the previous corollary and lemma.   ∎

REFERENCES

C1. Church, A., An unsolvable problem of elementary number theory, *Amer. J. Math.* **58**, 345–363 (1936).

D1. Davis, M., "Computability and Unsolvability." McGraw-Hill, New York, 1958.

D2. Dunham, C., A candidate for the simplest uncomputable function (letter), *Commun. Assoc. Comput. Machinery* **8** (4), 201 (1965).

G1. Gödel, K., Über formal unentscheidbare Sätze der Principia Mathematica und verwandter Systeme I, *Monatsch. Math. Phys.* **38**, 173–198 (1931).

H1. Hermes, H., "Enumerability, Decidability, Computability." Academic Press, New York, 1965.

K1. Kleene, S. C., "Introduction to Metamathematics." Van Nostrand-Reinhold, Princeton, New Jersey, 1952.

M1. Markov, A. A., "The Theory of Algorithms" (English transl.). Nat. Sci. Foundation, Washington, D.C., 1961.

P1. Post, E. L., Formal reductions of the general combinatorial decision problem, *Amer. J. Math.* **65**, 197–215 (1943).

P2. Post, E. L., Recursively enumerable sets of positive integers and their decision problems, *Bull. Amer. Math. Soc.* **50**, 284–316 (1944).

P3. Post, E. L., Recursive unsolvability of a problem by Thue, *J. Symbolic Logic* **12**, 1–11 (1947).

Q1. Quine, W. V., Concatenation as a basis for arithmetric, *J. Symbolic Logic* **11**, 105–114 (1946).

R1. Rado, T., On a simple source for non-computable functions, *Proc. Symp. Math. Theory of Automata*, pp. 75–81 (1962).

R2. Rosenbloom, P. C., "The Elements of Mathematical Logic." Dover, New York, 1950.

S1.   Smullyan, R. M., "Theory of Formal Systems" (Ann. Math. Studies, No. 47), Princeton Univ. Press, Princeton, New Jersey, 1961.

T1.   Turing, A. M., On computable numbers, with an application to the Entscheidungsproblem, *Proc. London Math. Soc.* [2] **42**, 230–265 (1936); **43**, 544–546 (1937).

W1.   Wang, H., A variant to Turing's theory of computing machines, *J. Assoc. Comput. Machinery* **4**, 63–92 (1957).

INDEX

U

W

Y